The R.A.M.S. Library of Alchemy

Volume 23

18 Short Tracts

Collected by
Hans W. Nintzel

R.A.M.S. Publishing Company

18 Short Tracts

Collected by
Hans W. Nintzel

Produced by

Restorers of Alchemical Manuscripts Society

R.A.M.S. Publishing Company

R.A.M.S. Publishing Company
117 Rutherford Lane
Stuarts Draft VA 24477

18 Short Tracts
Copyright © 2015 R.A.M.S. Publishing Company

First Edition 2015

ISBN-13 **978-1511516204**
ISBN-10 **1511516208**

Image Processing by Philip N. Wheeler

Printed in the United States of America

Table of Contents

Dedicated to Hans W. Nintzel,
American Alchemist
and
Founder of the
Restorers of Alchemical Manuscripts Society
(R.A.M.S.)

Disclaimer

Liability: The publisher does not warrant or assume any legal liability or responsibility for the accuracy, completeness, or usefulness of any information, apparatus, product, or process disclosed. The publisher makes no representation as to the accuracy or completeness of the contents of this book and specifically disclaims any implied warranty of merchantability or fitness for a particular purpose. No warranty may be created or extended by written sales materials or sales representatives. You should obtain professional consultation where appropriate. The publisher shall not be liable for any loss of profit or other commercial or personal damages, including but not limited to special, incidental, consequential, or other damages.

Introduction

Philip N. Wheeler

Hans Nintzel collected these small works together since each in itself is too short to require a dedicated book. Nevertheless, Hans considered them to hold valuable clues to the achievement of the Great Work in Alchemy.

I have moved two of the original works, Zoroasters Cave (Extract) and A Treatise Written by a Celebrated Philosopher, into The R.A.M.S. Library of Alchemy's Bacstrom's Notebooks Part 3, since their English translations originated in the Bacstrom notebooks.

The works in this Volume include some of the writings of Raymond Lully, Eugeneous Philaletha, Lamspring, Louis Grassot, Count Trevisan, and many short manuscripts on Alchemical processes from the Sloane Manuscripts Collection of the British Museum.

EXPERIMENTS for the PREPARATION

Of the

SOPHICK MERCURY

By Luna and the Antimonial Stellar Regulus of Mars

For the

PHILOSOPHERS STONE

Written by: Eugeneous Philaletha

An

Englishman and a Cosmopolite

Preparation of the Sophic Mercury

1. The secret of the Philosophick Arsenick.
I took one part of the Fiery Dragon, and of the Magnetical Body two parts, I prepared them together by a strong Fire, and in the first fusion there was made about eight ounces of the true Arsenick.

2. The secret of preparing the Mercury with his Arsenick, for the separating its Faeces.
I did take one part of the best Arsenick, and I made a marriage with two parts of the Virgin Diana into one Body; I ground it very fine, and with this I have prepared my Mercury, working them all together in heat, until that they were most exquisitely incorporated: then I purged it with the Salt of Urine, that the Faeces did separate, which I put away.

3. The Purification of the Sophick Mercury.
The Mercury thus prepared, is yet infected with an external uncleanness, wherefore distill it three or four times in its proper Alembick, with its Steel Curcurbit, then wash it with the Salt of Urine,

until that it be clear and bright, and in its motion leaves no tail behind it.

4. Another most elegant Purgation.

Take of decrepit Salt, and of the Scoria of Mars, of each ten ounces, of prepared Mercury one ounce and a half, grind the Salt and the Scorias very fine together, in a Marble Mortar; the put in the Mercury, and grind it with Vinegar, so long until no more of the Mercury appears; put it into a Glass Body, and distill it by Sand in a Glass Alembick, until all the Mercury be Ascended, pure, clear, and splendidly bright; reiterate this three times, and you will have the Mercury excellently well prepared for the Magistery.

5. The secret of the Just Preparation of the Sophick Mercury.

Every single preparation of the Mercury with its Arsenick is one Eagle, the Feathers of the Eagle being purged from their Crowlike blackness, make it to fly the seventh flight, and it is prepared even until the tenth flight.

6. The secret of the Sophick Mercury.

I have taken the proper quantity of the Mercury, and I mixed it with its true Arsenick, to wit, about four ounces of Mercury, and I made a thin commixed consistence; I purged it after a due manner, and I distilled it, and I had a pure Body of Lune, whence I knew that I had rightly prepared it; afterwards I added to its weight of Arsenick, and I increased its former weight of Mercury, in so much that the Mercury might prevail to a thin flux, and so I purged it, to the wasting of the blackness almost to a Lunary whiteness; then I took half an ounce of the Arsenick of which I made a due marriage, I added it to this betrothed Mercury, and there was made a temperature like Potters Loam, but a little thinner; I purged it again, after a due manner, the Purgation was labourous and a long time; I made it with Salt of Urine, which I have found to be the best in this work.

7. Another Purgation, but yet better.

I have found out a better way of purging it, with Vinegar and pure Sea-Salt, so that in the space of half a day I can prepare one Eagle; I made the first Eagle to fly, and Diana is left, with a little Tincture of Brass; I began the second Eagle by removing the superfluities, and then I made it fly, and again the Doves of Diana are left, with the Tincture of Brass; I conjoined the third Eagle, and purged the superfluities, by removing them, even to a whiteness, then I made it fly, and there was left a great part of Brass, with the Doves of Diana; then

I made it fly twice by itself, to the whole extraction of all the Body; then I joined the fourth Eagle, by adding more and more of its own humour by degrees, and there was made a very temperate consistence, in which there was no Hydrops, (or superfluous moisture) as there was in the three former Eagles.

8. I have found the best way of preparing the Sophick Mercury Viz. such as follows.

The Amalgamated Mass, espoused or joined very intimately by a due Marriage, I put into a Crucible, and into a Furnace of Sand for half an hour, but so that it might not sublime; then I take it out, and strongly grind it; then I put it again in the Crucible, and in the Furnace, and after a quarter of an hour or thereabouts, I grind it again, and make the Mortar hot, by this means the Amalgama begins to be clean, and to cast forth a great deal of Powder; then I put it in the Crucible again, and to the Fire as before, for a convenient time, so that it be not sublimed, otherwise the greater the Fire is, the better it is; so continually putting it in the fire and continually grinding it, till almost all the powder doth wholly disappear, then I wash it, and the Faeces are easily cast out, and the Amalgama becomes intire without any Heterogeneity; then I wash it with Salt, and again do heat it and grind it; this I repeat to the full cleansing it from all manner of Faeces.

9. A Threefold trial of the goodness of the Prepared Mercury.

Take thy Mercury prepared with its Arsenick of seven, eight, nine or ten Eagles, put it into a Phial, and thou shalt lute it with the Lutum sapientiae; place it in a Furnace of Sand, and let it stand in an heat of Sublimation, so that it may ascend and descend in the Glass, until it be coagulated a little thicker than Butter; continue it unto a perfect Coagulation, until it be as white as silver.

10. Another Trial.

If by shaking it in a Glass with the Salt of Urine, it be turned into an impalpable white powder of its own accord, so that it doth not appear as Mercury, and of its own accord in an hot and dry place it coagulates again into a thin Mercury it is enough; but yet better if being agitated in Funtain-water, it runs into small heads or particles, for if the grain be in the Body, it will not be thus converted and separated into small minute parts.

11.   The Third Trial.

Distill it in a Glass Alembick, from a Glass Curcurbit; if it passes over and leaves nothing behind it, it is a good Mineral Water.

12.   The extraction of the Sulphur from the living Mercury by Separation.

Take thy mixed corporal and spiritual Compound, the Body of which is coagulated of the volatile by digestion, and separate the Mercury from its Sulphur by a glass still, and thou shalt have a white Luna fixed and resisting Aqua fortis, and more ponderous than common Silver.

13.   The Magical Sol out of this Luna.

Out of this white Sulphur by Fire thou shalt have a yellow Sulphur, by a manual Operation, which Sol is the red Lead of the Philosophers.

14.   Out of this Sulphur, Aurum Potabile.

Thou mayst turn this yellow Sulphur into an Oil as red as Blood, by circulating it with the Volatile-Mercurial-Philosophical Menstruum, so thou shalt have an admirable Panacea, or Universal Medicine.

15.   The gross conjunction of the Menstruum with its Sulphur, for the formation of the Offspring of the Fire.

Take of thy purged, best prepared and choicest
Mercury, of seven, eight, nine, or at most ten
Eagles, mix it with the prepared Laton, or its red
Sulphur, that is to say, two parts of the Water, or
at most three, with one of the pure Sulphur, ground
and purged; N.B. but it is better that thou takest
two parts to one.

16. The working of the mixture by a manual Operation.

This thy mixture thou shalt grind very well upon a
Marble, then thou shalt wash it with Vinegar and Sal
Armoniac, until it hath put off all its black
Faeces; then thou shalt wash off all its saltness
and acrimony with clear Fountain-water, then thou
shalt dry it upon clean white paper, by turning of
it from place to place with the point of a knife,
even unto an exquisite dryness.

17. The putting in of the Foetus into the Philosophical Egg.

Now thy mixture being dryed, put it into an Oval
Glass, of the best and most transparent Glass, of
the bigness of an Hens Egg; in such a Glass let not
thy Matter exceed two ounces, seal it Hermetically.

18.   **The Government of the Fire.**

Then you must have a Furnace built in which you may keep an immortal Fire; in it you shall make an heat of Sand of the first degree, in which the dew of our Compound may be elevated and circulated continually day and night, without any intermission, & etc. And in such Fire the Body will die, and the Spirit will be renewed, and at length the Soul will br glorified and united with a new immortal and incorruptible Body: Thus is made a new Heaven.

FINIS.

THE FIRST MATTER

(No Author or Date)

When we turn our thoughts to the preparation of the Universal Matter, the highest Art of the Wise, in which is concentrated all the strength of heaven and earth, we must know that it is in all things, know what the universal Subject is, and how it is constituted; the fire of nature in its two diverse properties, the corporeal and the universal working power, which the philosophers call the First Matter. The foundation being the Mercury of the Wise also called the universal matter. But the Wise know that it cannot now be found in the world as one Thing, in which the 2 central fires of nature remain limited and corporeal. Therefore they say, that whoever can work through one Thing errs and does not follow the Art, but he who knows the Universal Matter has the true Wisdom, and for him the door of Wisdom stands open.

What God has created is found in the three Kingdoms, so we must earnestly look for this Universal Matter in them, and it is easy to find if we think in which creation the cleanest and purest Light of Nature of God can be found.

The Universal Subject is the most noble of God's Creation, in which God Himself burns for love.

When we look around in the whole of nature, we see that Man is God's most excellent work, for all creatures belong to the natural and perishable world, but Man stands in a two-fold position, for he is natural and also supernatural, held dearest by God and created over all things.

Though his body is perishable through the Fall by corruption and participating in it, yet his soul is born of God, supernatural and imperishable, for the Spirit of God has his place and home in him, and does not belong to the great world, but Eternity.

Therefore Basilius Valentinus says: Man is through his soul a divine entity, even though his outward body perishes even as the animals, but man can overcome this natural being through the Q.E. for he is the Centrum Centrorum and the Centrum Concentratum, as the small world is called, and Basilius says further, everything is found in the small that is found in the great one, namely heaven and earth, the elements and all that is in the Firmament.

We see too that the Wise, not without reason, say, "In you lies the most precious of all precious things, in man is found all the strength of heaven and earth. The Philosophers say, "Take on the Art, where Nature leaves off", for they wish people to understand Creation and how Creation came about. The philosophers after their magician Stilo say: This Universal Matter is the philosophic Mercury from which all things are taken and held, and man being the Centrum Centrorum understands the properties of all Creatures, and has the powers of the three Kingdoms of the lower world in triplicate in himself. He must realize that the beginning of the Light of Nature can be found in himself, for he is grounded in the eternal nature.

In order that we should not be confused by the various names of the Prima Materia, Sendivogious warns us truly that we must comprehend the hidden meaning, not simply the outward words, for that is their intention. Where Nature finishes in the Metallic kingdom and in our eyes has a substantial body, there we must start our Art. The Mercury of the Wise must necessarily be taken from man, as Riplaus shows when he writes, Man is the noblest of all creatures in the false creation, and in him are the four elements, one in another, which become one working, and in the mercurial family are many materials, which through the Art can be drawn out.

As the two central fires of nature lie in the Microcosmo, so are they also in the Macrocosmo, and in specie in the Spiritual fire. In the fire is an Astral Spirit, as a subtle nitre and is endowed with a living soul, which is the Archaus of the Motor

Universalis, that is in all things and which opens its most inner Centrum and pours into it a fermenting Property through which life awakens, and is brought to further use: Without this astral spirit the two corporeal fires are dead and not working, but when they are opened by the fermenting property they are made volatile, and all life awakens. Nothing can exist without this astral Spirit which is the life of all things, and in Alchemie nothing new can be born without it, or be placed in the Plusquam perfection, for the Spirit alone gives life to all. As all creatures have it for their beginning and in Gods order are held by it and nourished, so it is no stranger to any creature, for though it is Spirit, yet it can be in a body. Hermes says: The upper open Spirit is most hidden in the earth. Work the upper open Spirit with the lower secret one, so will the living one awaken the dead, and be to it as a Well of Life and work great wonders.

The Holy Spirit is indeed the agent which awakens and subtilizes the two corporeal central fires. When these two are brought to full perfection and their Universal Working power made substantial, the philosophical saying is justified:

"He who makes the volatile fixed, and the fixed volatile, understands the working of our Universal and Particular Recipe".

This is confirmed by Basilius in his Tract on Universal and Particular Things when he writes: "The Universal is the most precious thing of the Wisdom, and the three properties are one property, and is found and drawn out from one Property from which all metals can be made into One, and is the true Spirit Mercury and Anima Sulphur, which cleanses the holy salt, sealed under one heaven, living in one body and is the Dragon and the Eagle. The King and the Lion is the Spirit and the Light Word of Spirit, the body of the Sun colour becoming a medicine. Finally all three principles are one expression in the Love of God."

This Universality of the Stone is confirmed by the many born from the universal centre. Other Recipes are only used in particular instances and must be so observed.

That the Universal Materia lies only in the Microcosms is further shewn by Basilius in Tr. de Microcosmo. The natural means and remedy is found in the Microcosmos, metals and minerals follow after, and if one cannot find out the remedy, place one similar upon another similar and healing will result. It is also said, though it may be easy to make the Stone or Lapidum it is far harder to full comprehend it. All Philosophers including Adam, Soloman, Hermes and Theophrastus, by they ever so wise, recognize this difficulty. Dionysius Zacharias also realises that God placed a Divine working power in the Universal Stone.

The Wise always mentions the necessity of knowing how to prepare their Mercury in their writings, for they say "Our Spirit which enlivens the bodies of metals is a natural fire, declared openly by man or revealed by Gods Spirit. Flamellus fearing to offend God did not wish to disclose the Key to the portals of nature, by turning over the lowest earth to the highest heaven. Sendivogius says: "Though in a sense much is said openly nevertheless, the extraction of our salt or Mercury Sophici and that which bridges them over our water, is never openly mentioned but only revealed by God."

When the philosophers mention their salt or mercury, their WATER is simply a matter of speech. They speak of their water as a dry water which does not cleanse the hand. What they really mean by this is that the salt is the Key to the Art, for in the SALT lies the opening of all things, and in such lies the Universal Menstruum of the white Mercury. Of this Mercury the philosophers say: It is a Stone and yet not a Stone, when a fiery air salt goes into all bodies and opens them. It is a Stone however when it opens and unites with the body, then it belongs to the fiery grade of nature, may become dry and be processed. Then is the white mercury a salt which has no resemblance to the ordinary salt, but which has an Alkali and an Astringent substance, of which Philaletha says, "Our mercury is to be found nowhere over the earth, but it is a Son which is given to us by a wonderful Art."

Its beginning is partly heavenly, partly false, for it must give birth to a Light Spirit, in it are two central spiritual fires, these at the beginning of all Procreation covered over the waters, acting according to Gods Hand, and until a certain hour gave all creatures nourishment and multiplicity. It is false however through its false being, overcomes love and its existence and thereby partakes of both the lower and higher powers. From its heavenly source it draws all the powers of heaven, and lives in the Air, therefore called the Sea of the Wise, and Basilius again says. The Recipe of all metals is a heavenly volatile watery spirit, which is contained in the AIR and in the earthy kingdom and WATER, seeks its natural home and the air is the secret and hidden earth, which hovers over our heads and from which through the elementary environments is drawn our white nitre. Paracelsus writes of this, "One must know the anatomy of the living body, and from it draw the Essence (without hurting the body) from which all the wonders of the world can be produced". This would be the great secret of the Adepts which they keep concealed, and Paracelsus shows that the two central fires should not be taken roughly from people but gently when overflowing of itself.

Ur in Caltha has in it the cold central fire which stands in its noble salt, and that is our Magnet, Philaletha calling it Moon and Copper. In the red that is in the Stercus Adamish earth is the Electrum minerale immaturum, the Electrum electissimum, the secret white saturnine minera, still in its water, and in whose strength all metals lie hidden, and from which can come forth minerals like Antimony, Iron, Solis Gold, the sulphur of the philosophic mars, iron and the gold in which the warm central fire or our acidie sulphur, through which our astral mercury receives it. For this sulphur is the clearest fire and greatest Balm of Nature.

Sendivogius warns us that we should not be led astray by these above mentioned minerals, for he says: "One must not look for the warm central fire in the ordinary metals of common man, for the gold in ordinary mortals is dead, but ours is alive and has a spirit, it is this which one must take."

This upper astral spirit Sendivogius calls our water or Dew, from which our nitre is drawn, the mother of the centrum solis and moon. Hermes says: The sun is its father, the Moon its mother, the wind carries it in its belly. It is called by many names, the Subjectum, Hyle, the First Matter of the world, for this Spirit is the same light Power hovering in the beginning over the water, the Archaus, the first working nature or Beginning of all things.

Protheus also, for he takes all manner of creatures to himself. The Saturnian Spirit, for he brings all things before us. This astral spirit through its incessant working and motion imparts its strength to the lower world. For if this spirit ceased for a moment, then would all creatures die and decay.

> God the Almighty Creator of Heaven
> The earth lives in Light
> The Light in Spirit
> The Spirit in Salt
> The Salt in the Fire
> The Air in Man
> The Man makes visable the Air and Spirit.

Oh single eternal Heavenly and true Life, the salt and tree Jesus Christ! This fruit, Apocaly xx11 v. 2. which leads and holds us in eternal life, I pray you that we may confound the false Tree of Life, Gen. 2. v. 9:13 & 22 and hand over the Walls of Paradise these pages of prose. May we in this life reach our goal for the refreshment of our mortal bodies, to the confounding of death by

practising the true knowledge and working of our
Art!

AMEN.

THE GLORY OF LIGHT

B.M. Sloane 2219[1]

translated by: L.F.P.

The Truth seems buried because it brings forth little Fruit, but it is great and prevaileth to make all things manifest, so far as is possible to all men; for in common sense and reason all agree in Mysteries never so that none may speak of Science without Knowledge, which breaks the Gates of Brass, and cuts asunder the Bars of Iron, before the Eyes of Understanding, that the Treasures of Darkness may be opened, and the bright and fiery Sword discovered, which turns every way to keep transgressors out of Paradise; for if we consider wherein ye Celestial and Terrestrial Bodies agree, we shall find something objective in the Inferiour Bodies, whereby they communicate their Celestial Virtues and Influence, which precident Art doth imitate, to produce a Glorious Substance of commixed forms and of clearest Virtues and Beauty beyond expression.

The Mathematicians say the Celestial Influences do hold and govern every natural Body, and by many

---

[1] This work is known to be from 1753 or earlier.

Unities collect a quantity subsisting without Shadow; for the real Virtues effect to be specificate, and so Living Fire gives Life to other things, which Central Substance of Celestial Virtue or Form of Metals, is the Subject of this short Discourse.

That Urim and Thummim were given in the Mount cannot be proved; That they are potential from the Creation may appear, for they were Substances whose Name and Essence did predicate each other being convertible terms, the Name and Essence, one. The Words signify Light and perfection, Knowledge and Holiness, also Manifestation and Truth, even as Science and Essence make one perfection. It is likely they were before the Law given, for the Almighty commanded Noah to make a Clear Light in the Ark, which some take for a Window; others for the arching and bowing of the upper deck a cubit. But saith the Text, "Day and Night shall not more cease", (Gen: 8:22) it seems it did not then cease. And whether there were one or many Windows is uncertain: But when the Windows of Heaven were opened and the Air darkened by pouring out Rain, the Sun not giving his Light, but prohibited the Generative Spirit of the Creatures in the Ark, what exterior Clearness could be expected.

Therefore some of the Rabbins say, the Hebrew word ZOHAR, which the Chaldee translate NEHER is not found in the Scripture but in this place, so that like the word, it seemeth to be a Rare Light, and that which is generally doubted to be, the Creator commanded Noah to make by Art. Other Hebrew doctors say, it was a precious Stone hanging in the middest of the Ark, which gave Light to all Living Creatures therein: This the greatest Carbuncle could not do, nor any precious Stone that is only natural;But the Universal Spirit fixed in a transparent Body shines like the Sun in glory and gives sufficient Light to all the room to read by: Therefore it is most probable this was the Light that God commanded Noah to make to give Light to all living Creatures, for it is of perpetual durance: And whereas Tubal—Cain is said to be a perfect Master of every Artifice in Brass and Iron which some hold to contain the whole and perfect Decoction of the Metallical Virtue, wherein the Central Virtue is most abundant, and makes the happy more admired who walk in the midst of the Stones of Fire; For where there are two things of one Nature, the chief is to be understood, therefore the mention of Fires, Pure Fire is preferred. The Scarlet Veil in the Temple seemed ever moving, and signified Pure Fire, generative and moving, which fixed in clear bodies is Urim and Thummim.

Although Essences are not without great difficulty made manifest by themselves, yet the clear Vision thereof makes the possibility unquestionable. As at Elisha his Prayer, his Servant saw the Horses and Chariots of Fire about his Master, which before he saw not, so are these apparent when the Invisible is made Visible. Some think that Urim and Thummim were not Artificial, because they are said in the Text to be put into the Breastplate but not to be made (Exod. 28:30). But this point may be cleared by observing the several kinds of making as betwixt those things made with Hands, and those things which are only made visible by Effect, for where Nature and habitual Virtue do meet together the perfection is more absolute by a new generation, as the pure Sulphur of Metal by an inward Power doth purify itself by Ebulition, not by the first and remote Causes, but by the second and nearer, whereof the Philosopher saith, "The Secret of all Secrets is such a disposition which cannot be perfected with Hands, for it is a Transmutation of Natural Things from one thing to another". Also it is said, "The Artist taketh impure Spirits, and by Sublimation, Nature and Art cleanses them into Bodies, Pure and Fixed; so that the Bodily nature doth Eternally predominate, and being more than perfect, doth give perfection to other things.

Now that these perfections have their beginnings from two Lights, both the Text and the Ancient Philosopher make plain: But ignorance and the Matter of the Elements are the Iron Gates which must be cut in pieces before the Invisible be made Visible. For the Natural Urim and Thummim, the Philosophers affirm what they have seen and done, and that they did nothing, save that they know before: So that a perfect Knowledge is especially requisite to make a perfect Art. Therefore we are now to consider the means to attain this End. The Lord gave Bezabel Wisdom, Understanding and Knowledge, these are the Means; for Gold is dissolved by Wisdom; in Contrition, Assation, and Fire. The End is directed to invent works in Gold, Silver, and Brass, which is not to be understood according to the sound of the words, but according to the intent of all Distillation, to extract the Inward part, and manifest the Central Virtue: for where the perfection of the Matter is glorious, the perfection of the Form must be more glorious. The Sun and Moon are as the Parents of all Inferiour Bodies, and those things which come nearest in Virtue and Temperature are more excellent. The Suns motion and Virtue doth vivify all Inferior Bodies, and the pure form of the Terrestrial Sun is said to be all Fire, and therewith doth the Celestial Sun communicate most Virtue: Therefore the incorrupted quality of pure Sulphur being digested in external Heat, hath also regal Power over all Inferiour Bodies. For the Sun doth infuse his Influence into all things but especially into Gold; and these Natural Bodies do never shew forth their Virtues until they be made Spiritual. One of the Rabbins saith, "They made in the Second Temple Urim and Thummim, to the end they might make up all the eight Ornaments, although they did not enquire by them, because the Holy Ghost was not there. And every Priest that spake not by the Holy Ghost and on whom the Divine Majesty resteth not, they enquire not by him." So it is with Sacramental Bread which hath no significance before Consecration: But these men had the Spirit of Bezaliel, and made the Natural Spiritual Bodies, which Soverign Tincture some say so purifieth and causeth the Radical Humour so to abound, that the Children of the fourth Generation (yea, some say the 10th) shall perceive the effect of such present Health of their Ancestors.

The two Staves which uphold man's Life is native Heat, and Radical Moisture, which requireth all care to observe equal proportion and mixture; like a Lamp, where neither the Flame nor the Oil must surpass, lest the Oil exhaust or the Flame suffocate, for there is a possibility and aptness in

Nature to attain Eternity, seeing natural desires are never altogether frustrate. And this aptness extended itself to immortality as it was before the Fall and shall be after the Resurrection but there is one form of Nature appointed after the Fall, and another by Corruption of Parents, for there are perfect terminative and privative ends. The Hart and Eagle renew their Youth so that it is possible for a Man to obtain that which is not denied to unreasonable creatures.

Philosophers say that if you have once finished this Work and should live one thousand years ye might give what you will and when you will, without danger of diminition, as a Man that hath Fire may give to his Neighbour without hurt to himself. Marcus Varro saith, "There was much more Mystery in the Flammine Ceremonies than one understood: Vesta signified pure Earth and Fire Internal, of whom it is said, Vesta is Earth and Fire. Earth undergoeth the name and so doth fixt. Vesta is both."

"Thus is shown forth in a Work by Fire,
The Mighty Vesta and her pure attire."

Philosophy is nothing else but the study of Wisdom considered in a Created Nature, as well subject to Sense as invisible, and consequently Material. And Wisdoms Central Body is the Shadow of Wisdoms Central Essence, and the Moral interpretation can never exclude the real Effects from ocular demonstration, but where Reason hath Experience, Faith hath not Merit, and without Faith there is no knowledge of any excellent thing, for the end of Faith is Understanding.

The Rabbins hold every natural beginning to be either Matter, or the Cause of Matter, viz. the four Elements; Others are of opinion the Creator first made one Pure Matter of which He made the Four Elements. But here beginnings must be well understood, for there are beginnings of Preparation, and beginnings of Composition and Operation: for the Artist was commanded to devise works in Gold. That is, from the Object to the possibility.

For if the Matter be Glorious, the Form must be more Glorious and though the Spiritual Nature be more operative, yet the Bodily Nature must predominate Eternally.

So that to make the Corporeal Spiritual, and the Spiritual Corporeal is the whole scope of this intention. Yet the Spiritual is not the first, but the Natural, for Corruption must put on Incorruption and Mortality, Immortality. For that which is of greatest duration and most Abundant in Virtue, doth most excel in Glory and Beauty, and so fittest to make Urim and Thummim. For Power and Honour are in this Sanctury. But because the greatest things are not done by strength or habit of fingering as also because the Intellect doth so far excell the Sense this is a Work of second Intention, and the beginning upon the Virtue of Elements. That is a pure, bright, and clear Water of Putrefaction, for the perfection of every Art, (properly so called) requires a New Birth, as that which is sowed is not quickened except it die. But here Death is taken for Mutation, not for rotting under the Clods. Now therefore we must take the Key of Art and consider that the secret of everything is the Life thereof; Life is a Vapour, and in Vapour is placed the Wonder of Art, whosoever hath Heat agitating and moving in itself by the internal Transmutation, is said to Live.

This Life the Artist seeks to destroy and restore an Eternal Life with Glory and Beauty. This Vapour is called the Vegetable Spirit because it is of degree of heat with the hottest vegetable, and being decocted until it shine like brightest Steel, ye shall see great and marvellous Secrets, not by

Separation of Elements by themselves, but by predomination and victory of that Pure Fire which like the Celestial Sun, enters not materially, but by help of elemental Fire sends forth his Influence and Impression of Form. Here we must observe difference of perfections, for although ye have now the Fountain of Life, and Centre of the Heart, the Universal Spirits which lives in the Radical Humidity and doth naturally vivificate, and is the Masculine Seed of ye Celestial Sol. Here is that rule made good, except you sow Gold in Gold ye do nothing. Therefore we must take heed what we understand by Gold, whereof there are three Sorts, Vulgar, Chemical, and Divine, which is therefore so called because it is a Spiritual Gift of God.

The Theosophists are persuaded, by exact Diet, and by certain forms of Prayers at certain Times to obtain the Angel of the Sun to be their Guide and Director; The Philosophers advise to take the like Matter above Earth that Nature hath under earth. Others say that the most precious Treasure riseth from a Vile thing, all which are easily agreed if rightly understood, for in the lines following the same Author saith, "The Vile thing is from the Sperm of Gold cast in the Matrix of Mercury by a prime Conjunction". Others affirm Azoth and Ignis to be sufficient for this high perfection, the which Azoth among the Germans is Silver, with the Macedonians Iron, with the Greeks Mercury, with the Hebrews Tin, with the Tartars Brass, with the Arabians Saturn, and with the Indians it is taken for Gold. All which being adverse in Nature, are potential in one Composition, and by the Dual of Spirits the Celestial Gold obtaineth Victory over all the rest, and is made (though not with hands) a Body, Shining like the Sun in Glory, which is called Ens omnis privationis expers or Thummim. This is the King that made the pure, clear, bright Fountain, and of it was made himself. The fair Woman, so Loving the Red Man she became one with him, and yieldeth him all Glory, who by His Regal Power and Sovereign quality reigneth over the Fourfold Nature, Eternally. But if any shall understand either Common or Chemical Gold to be the Subject of this Sacred Body, he is much mistaken; for a Glorious Spirit will not appear save in a Body of his own kind. Although pure (Manchet) be made of the finest Meal, yet Wheat is not excluded, and so Bread is most properly said to be of the second and nearer Causes, rather than of the Remote, notwithstanding that which is made with hands.

After we fell from Unity we groan under the burden of Division; but Three makes up the Union, first temporary, and after Eternally fixed. He that knows a thing fully must know what it was, is, and shall be. So to know ye several parts of a successive Course is not a small thing, neither the honour little.

In the right use of the Creature: Air turned into Water, by his proper mixture becomes Wood, and the same Wood by water is turned into Stone. As a Spring in Italy called Clitummus makes oxen White that drink of it. And a Water in Boetia makes Sheep Black that drink of it. And the river in Hungary turns Iron into Copper. What excellency things may attain by habitual Virtue, or what Power, when Nature and Art make one Perfection, who is able to express.

What reverend Martin Johnnes Rupicisa affirmeth, "The exalted Quintessence upon the breaking of the Glass sendeth forth such fragrant Scents, that it doth not only delight those that enter into the house, but even Birds that fly by will sit on the window sill so ravished with delight, that you may take them with your hand." And if you desire by Art to have a thing of admirable sweetness and odour, you will take a Subject of like quality to exalt into such excellency. (Beza made ye Perfume).

The proper quality of Fire and Air is Sweetness, it is but approximate in Earth and Water, what Bodies shall we find where these are most abundant to be wrought upon.

As the Celestial Bodies work qualities in other things, yet have none in themselves. So the Metallical Bodies give no Tincture: yet are most abundant in Tincture. Air is Cause of Life, Mercury is a cocted Air, Aethereal and truly Homogeneal, which doth after a sort congeal and fire; It is called a Crude Gold, and Gold a fixt and Mature Mercury, and although the Crude quality be cold and dry, yet the Internal and Ethereal Spirit is held hot and dry, and some hold for the excellency of his Temperature that it is all Fire or like to it, whereby it is dissolved, howsoever, it is at large proved that these Bodies are most abundant in pure Fire and Air, whose proper quality is Sweetness. Therefore these are the fittest Subjects to make the most precious Purfume in the World, and considering clearness and brightness is the Centre of each thing. And these Bodies have both centre and superficies, clear and bright when they are purified by Art, and the Bodies made Spiritual and those Spirits corporate again, they must necessarily be Bodies of greatest or clearest Light and Perfection. As one compareth a Glorified Body to a clear Lantern

41

with a Taper in it, Saying, "The more a man excelleth in Virtue the greater or lesser was the Taper". But the work cannot be manifested without the destruction of the exterior Form, and restitution of a better, which is the glorious Substances of Urim and Thummim, which in their Being and Physical Use preserves the Temple of Man's Body incorruptible. Some observe not just difference betwixt Liquefaction and Solution but all corrosives of violent operations Nature hates, because there can be no true Generation but of like natures, because there can be no true Generation but of like natures, neither can you have the precious sperms without Father and Mother, and although One Vessel is sufficient to perfect the Infant in the Womb, yet Nature hath provided several Breasts to nourish it and different means to exalt it, to the strength of a Man.

How Gold should be burnt, which the Fire cannot consume, is questionable, but every Exaltation of this Sovereign Spirit adds Tenfold Virtue and Power, then take one part of this Spirit, which is become as insenseable Dust, and cast upon Molten Gold, it turns all into Powder, which being drunk in White wine openeth the Understanding, increaseth Wisdom, and strengtheneth the Memory. For here is the Vein of Understanding, the Fountain of Wisdom, and the

River of Knowledge. The Truth of everything is said to be in his Incorrupted Nature, for nothing shall rest Eternally visible at the last Fire but that which is of pure Virtue and Essential Purity.

Truth and Science is not led by chance or Fortune, but the Spirit of God guides by the hand of Reason. And it seems the Prophets approve of these Stones of Fire, some mentioning the Stone of Darkness, and as it were, Fire turned up, other the Stone of Sin. And Ezekiel the Stone of Fire attained by wisdom, which he differeth from the natural precious Stones as pure Fire from common Fire.

Therefore let modesty let that possible, (?) whereof he understands not the terminations and degrees, neither refuse the Waters of Shiloah because they go slowly, for they that wade in deep Waters cannot go fast.

To obtain the Treasures of Nature, ye must only follow Nature who gives not like time to every generation. But as the Mare has ten months, the Elephant hath three, or as some say, nine years, and fifty before conjunction. Be patient therefore in a Work of Nature, for thereunto only is promised victory, and the chief errors in Art are hastiness and dulness.

NOTA. Of the Substantial qualities, Sulphur, Salt and Mercury.

A substantial quality arising from the first Mixture of the Principles is Threefold:

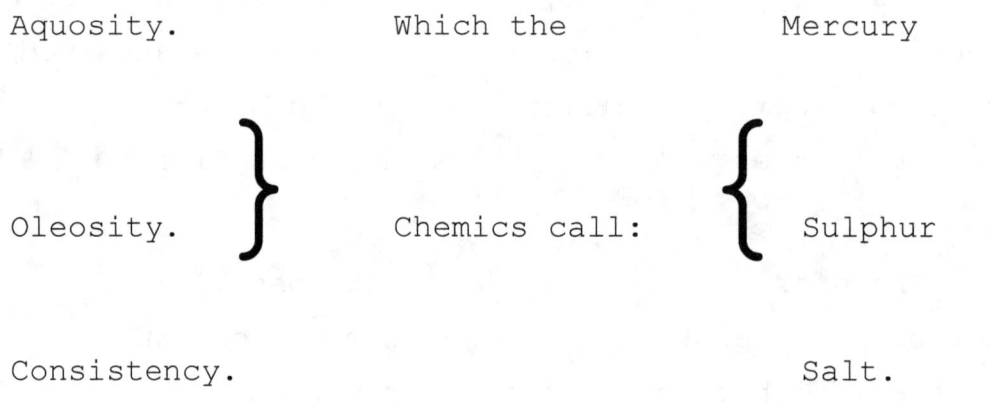

Aquosity.		Mercury
Oleosity.	Which the Chemics call:	Sulphur
Consistency.		Salt.

NOTE 1. These flow immediately from the combination of the first Principles.

FIRE.

Salt, Sulphur.

Spirit _____ Matter.

Mercury.

For as in the beginning the Spirit conjoined with the Matter produced the moving of the Waters; so Mercury is nothing but Motion, the first fluid thing which cannot be fixed nor contained within a limit; and Salt is Dry and Hot, and uncorruptible, just as Spirit and Fire, it is preserved by Fire, it is Dissolved with Water or Mercury, but turns neither to Flame nor Smoke, though it is a most Spiritual Creature, and every way incorruptible.

The Sulphur, what is it but Matter mixed with Fire? For why doth it delight in Flame, but that it is of a like nature, and in compound things it is the first thing combustible, or apt to be inflamed.

NOTE 2. But beware that you understand not our Vulgar Mineral Salt, Sulphur and Mercury, for Quicksilver, for these are mixed Bodies: Salt Earth, Sulphury Earth, Mercurial Water; (that is, matter wherein Salt, Sulphur, and Mercury are predominant, yet with other things adjoined, for Salt that parts apt to be inflamed, and Sulphur some Salt, and some Mercury, but the domination is from the chiefest).

These qualities cannot be seen as they are in themselves, but by imagination, but they are in all things, as Chemists demonstrate to the Eye. Who extracts crude and watery parts out of every Wood, Stone, etc. and other fat and oily parts, and that which remains is Salt, that is Ashes, so the thing itself speaks, that some liquor is Mercurious; (as Vulgar Water and Phlegm) other Sulphury; (as Oil and Spirit of Wine), others Salt and tart (as Aqua Fortis) also we find by experience in the benumbing Ashes of the Members, that some Vapours are crude, others sharp. God produced the qualities intrinsically that the Substance of every body might be formed. For:

> Mercury giveth unto things fluidity, Coition and Crudity, and from thence Incombustibility.

> Sulphur giveth unto things Softness, Cleaving together, Fatness, and from thence Inflammability.

> Salt giveth unto things Consistency Hardness, Aptness to Breathe, and from thence Incorruptibility.

That Mercury giveth fluidity and easy coition of the Matter appears out of Quicksilver, which by reason of the predominancy of Mercury is most fluid; So that it will not endure to be stopped or fixed. It is also most crude, so that it can neither be kindled nor burned: But if you put Fire to it, it flies away into Air. Now that the Coagulation of

Bodies is from Sulphur, as it were Glue, appears from thence that there is more Oil in dry, solid, and close Bodies than in moist Bodies; also because Ashes (after that the Sulphur is consumed with Fire) if you pour Water upon them cleave not together in a lump, but with Oil or Fat they cleave together. Now the Chemics extract Oil out of every Stone, leaving nothing but Lees, no part cleaving one to another any longer.

And that Salt gives consistency appears by the Bones of Living Creatures, out of which Chemics extract meer Salt, also all dense things leave behind them more Ashes, that is Salt. God therefore with great Council tempered these three qualities together in Bodies. For if Mercury were away the Matter would not flow together to the generation of things: If Salt, nothing would consist together or be fixed. If Sulphur, the consistency would be forced and yet apt to be dissipated.

Lastly, if there were not Sulphur in wood and some other Matters we could have no Fire, but Solar on the Earth (for nothing would be kindled) and then what great deserts would the Life of Man endure.

The Marriage of the Elements,

or How to Set Nature to Work.

It is not prophetical that all men shall wisely consider the Works of God to the end they may know how to value them rightly, and make just difference between corporeal and spiritual things, and corporate Spirits. For although Spirits possess not place, yet they fail not to fill every part by contact of their Virtue and in the use alters both quality and quantity. The perfect and distinct knowledge whereof doth necessarily manifest the thing sought after by the proper and appropriate qualities from the cause to the effect, openeth the internal Beauty of a true and natural Essence as plainly as by seeing that you see.

And also sheweth the terminative, privative, and perfect end of every particular Art, which is the richest of intellectual Treasures, because Science and Essence are One, and where the several Works and successive are apparent, the time need not be limited (like the men of Bethulia) for only at Elisha his prayer, his Servants Eyes were opened to see invisible things, which all which rise in Glory shall see. It was held of old nothing deserves the Love of an honest man save the internal Beauty, therefore they held Love or natural affection to be the first cause of Motion, like as the Heat and Motion of ye Sun and of the whole Heavens hath power in all things created under Heaven and by their influence and Radiation, all things increase, grow, live and are conserved and by their recess they mourn, wither and fall and droop, yet they do not necessitate any Art, their force being most on imperfect things, for a Body of equal temper receives little alteration from the Constellations, because the Earth received virtue before the Heavens were adorned with Sun, Moon and Stars. That is to be distinguished in Reason which is not distant in place, and different things in being, and in being and use.

For change of quality brought confusion, and a better change, renovation, the Historians affirm the River Nilus vapoureth not, by reason of the long decoction under the Sun, yet is the Water most wholesome and medicinable and the neighbour easily begins to encrease in weight the 17th. June (and not before) even then when the river begins to rise. Which sympathy of the distant Water and Earth by the power of Heaven is not against Nature, although beyond ordinary reach.

Therefore for a leading cast, let us observe the concord of Metallical Bodies, which like the first Male and Female have not several beginnings but are all from a Sulphureous Vapour, which by help of influence, instrument, digestion and a Masculine and Feminine Virtue, connection proper and appropriate qualities, they obtain their perfection by the power of God his Ordinance. Yet as every Earth yields not like Metal, so every Metal yieldeth not like Central Virtue. Therefore according to that Creating Command, everything should increase in his proper kind (not divers) and time makes the number infinite. The Ancients, reading the great Book of Nature, find no abridgement to assimulate the Mystery of Nature, save Man and the Stone, both which are called Living Stones, whose original mortification purity and difficult exaltation are of infinite virtue. They observe also a Celestial and Terrestrial Sun, which they parallel with man, because they are only capable of true temper, which is certainly possible, although seldom enjoyed; but to gain this precious Treasure of Life and Wealth, we must make sufficient provision, like men that deal with great persons, for Gold is Lord of Stones and noblest of Metals, and by his proper Regiment doth multiply himself infinitely, therefore Geber in his Book of Denudation saith, "In Gold are ten parts heat, ten parts humidity, ten parts siccity, which triple perfection makes an absolute Unity, Body, Soul and Spirit, being Eternally united, because Unity is a generical quality of all that is One, and is an effect of the Form which doth produce it. For of all kinds of governments, ten is the most perfect, and for the natural subsistence no composition is like unto Gold, for it is of most perfect temper and equal mixtion." The Miracle of Nature, A Celestial Star, A Terrestrial Sun, the Fountain of Life, The Center of the Heart, the Secret Virtue of all Celestial and Terrestrial Bodies, the Masculine and Universal Seed, first and most powerful of the Sulphureous nature, the great Secret of the Almighty. It hath most Form Entity so most virtue and operation. In it the Elements are Elementised, it is called Sulphur, and Sulphur Fire, yea, it is said to be all Fire, or like to that in which it is dissolved, and as Light is the Center of Heaven, and Soul of the World, so Brightness is the Center and Celestial Virtue, the form of Gold, whose admired mixtion nothing merely natural can dissolve, nor anything artificial, except it agree with it in Matter and Form, and do remain with it in the recongealation. This virtual influence, enters potentially and dwells in the Radical Humidity and no other thing, whether from Heaven or Earth, the heat, yet it is not visible before virtue be matched, for there is best concord where it is most abundant. But whither shall we mount to match this miracle of Nature. The Historians tell us of an Eternal Liquor of most strong co—action, rained down from Heaven, here is like descent, she is called Hypericon or daughter of the Sun, a body of like weight and virtue with Gold, fair, clear, quick, only cocted and brought from the Empire of Humidity, to suit the person, which in her crude nature shows strong affection and turns the noblest of the Metals into her own colour. Therefore the Artist studies how to disponsate these two, and first denudate the Lady of her Frosty Garments, that she may have the first activity, and liquify her fettered Lord, then are they both in the power of Art to better. But it is objected this heaven-born Hermaphrodite, Hermodactylus, or Hydromel, is of a nature so obstinate and incorrect, she will by no means receive the best impression. Consider, her namessignify mixt matter of contrary quality, therefore may be separated, and although her Original obscure her condition, because it is unknown, yet her virtues and qualities are known by the innate affections and subsistence, for it is an Airy Body, or Air itself with a Mercurial Spirit, subsisting of internal heat and external cold. Others say it is composed of the Spirits of the world corporate in the Womb of the Earth, and apt to receive the qualities and properties of all natural

things, as Wax impression, and being composed of
Spirits the weight is of greater wonder. Others say
it is a Crude Sperm not sufficiently decocted (yet
not to be prophaned). Others call it an Immature
Gold which kills itself, and the Father and Mother,
to bring forth a pure Infant. By her they overcome
the Fire, she is the perfection of the Universal
Medicine. What conformity, what similitude, what
identity she holds with her Metallical Nature, being
the Original Matter and Substance thereof, and may
be coagulated to the equal temper of Gold, is as
easily seen as the whiteness in Snow. Therefore the
Ancients magnify the most Blessed, who created such
a Substance and gave it such properties as no other
thing in Nature doth possess. Yet we see it is a
Viscous Matter, which hath superfluous humidity, and
proper and approximate qualities, separable and
inseparable accidents. Therefore the separable may
be removed, to which end she is included in a Well
of Tears that the Waterishness may be vapoured, or
through long decoction by dryness vanquished, then
it doth, as it were, congeal and Fire, and become
more apt for durance and extension, for whatsoever
is contrary to the natural property, doth
debilitate, and like by his like is nourished, but
heat is contrary to cold, and the natural property
of scalding heat is to weaken the dry, the fresh
Water adds power and heat, heat augmented becomes

Fire, and time turns strength into Corruption. This Glutionous Substance hath internal heat, from which is the Life and Death of the Elements. Therefore as common Fire bringeth all things to his own nature, so the external, working upon the internal heat, it doth necessarily obtain victory. Therefore if you can believe that heat and dryness shall overcome cold and moisture, that linear and successive course, hidden to all the World is open to you. Therefore as Nature delights in concord, so the Lovers and Searchers into Natures Works must be of constant minds, and Gedion—like resolve to raise the City Meros, refusing to assist the public good, and then to the Marriage for this Princess never unveiled her Virgin beauty except to him that hath skill and power to espouse her in a bed of Love which none can do before the dispoiliation of the exterior form, but the obstacles removed, and Nature set on work, the Eternal decree doth necessitate the effect, for being now warm and blithe, and apt for new generation, and pounded with her Lord grated to dust, his unnatural softness deceives the sense, and they passionately condole each others exile, and in their embraces fall into a swound, until their dissolved Bodies show Corruption, and the more pure being corrupted are more wild. Then the Artist finding them out of their Indian Paradise, collects their Central Virtue, and raising them from Death, leads them to the thorny path to threefold happiness, and by fiery trial, purifieth the quantity, and changeth the quality, and so bringeth them to perfect Rest, whereby they have power over the Bodies of Men and Metals, and are crowned in token of their dignity and boundless territory. Now concerning the rarity of true Knowledge, the honour and dignity of the thing desired, what Spirit is so ignoble to think much either of cost or time, when that which is sought is of all Terrestrial Treasures most excellent.

That the Regeneration of Man and the Purification of Metal, have like Degrees of Preparation and Operation to their Highest Perfection.

The first Beginnings of Transmutation or Naturation are the smallest measure of pure Sulphur, which hath Riches and Honour in the left hand, and Length of Days in her right.

In natural Generation the Form prepares the Matter, yet there are precedent preparations. The Beginnings of Transmutation must be distinguished.

Some are beginnings of Preparation, and some are beginnings of Composition. Beginnings of Preparation in the Well of Tears doth qualify the coldness and dullness of the crude disposition, and tame and subdue the fearful quality of swift flying, and changeth the colour of this Eternal Liquor, turning the inside outward, and adding heat by the Internal Sulphur of the Homogeneal Body, which is by means of changed Water, because Water by Water can only be extracted, yet is it excluded in the conclusion: for though it be a necessary preparation to the alteration following, yet it is but that servile passive which hath the first operation, being preserved unhurt in weight and purity.

Beginnings of Composition are those inward operations and changes that follow after the scalding Deluge, which by mixing with the fixed Sulphur, doth dissolve the stubbornness of his nature, and by help of the External heat, the Internal Sulphur is excited to operation, and purifyeth the Substance, but only to a Pale Whiteness, more hurtful than profitable to the Body of Man, what these are, shall afterwards appear.

Again, Sulphur must be distinguished, White Sulphur and Living or Reviving Sulphur. White Sulphur is of like operation, and is perfected by restraining, and healeth almost all diseases, and tingeth to White ad infinitum. By knowledge hereof, even meer natural Men have believed the Resurrection, became Sober, Chaste, Temperate and Patient, not doubting within ye Center of complete White, rests the Red Stone of most delight. This hath caused Men justly to condemn all Cementations, Calcinations and Citrinations, being enlightened with the glorious Object, which is as clear as a Crystal Looking Glass. Reviving Sulphur is the Secret of Secrets, and the glory of the whole World, and only proper to such, whom the Creator apted by way of natural disposition, for they do not only mortify, but purify a Pure Body, quickening it with the same Essential Form, and are said to make a Spiritual, because there is no corruption to resist the Spirit, but the Bodily nature being wholly subject, is with the Spirit Eternally fixed in a transparent Body, shining as the Sun. Therefore the Conclusion must be understood, of the second, and not of the first, for though a man have never so much White Sulphur, if he gave not of this Reviving Sulphur, he is as far from that precious Spirit which hath power over all Inferiour Bodies as any other, for only that which is of the nature of the Sun, shall shine like the Sun in Glory.

A synod of the Philosophers advises us, in seeking the Treasures of Health and Riches, we show our affections to virtue and prudence, like Soloman, asking wisdom, Riches was given as advantage.

Let us search then for Celestial Virtue, which is in the Centre of all things, so will it be manifest the Sovereign Spirit of Health and Riches, for the Vegetable Sulphur is the first Mover in Nature, and only the Mercurial Nature hath the power of Metallical Life and Death. Crude Mercury is originally a Vapour from clear Water and Air, of most strong Composition cocted, or Air itself with a Mercurial Spirit, by nature flying, Ethereal, Homogeneal, having the spirits of heat and cold, and by Exterior and Interior heat, doth congeal and fix. Also Gold is a fixed Fire or natural Sulphur or mature Mercury, and may be made more volatile than Mercury, but only by divers Mercuries made of Mercury is Nature set on work and the Fixed Body loosed, the Vegetable Sulphur created and the Universal Spirit fixed. How the Authority of Ancient Writers, Divine and Natural Reason doth assure us, that this, and no other is the true course to manifest and corporate those Lights wherein the Creator hath heaped up Virtue and Power.

But it is objected, The Philosophers do not agree amongst themselves.

Answer. Instruments of divers strings make sweet harmony if they be well tuned, but their readers do rather seek to over-rule them, than by painful industry to find their Consent.

Objection. They affirm contraries.

Answer. The Artists intention is to agree contraries.

Objection. They exclude Gold and Mercury from the creation of the Stone.

Answer. Because their Crude Matter is from the destruction of the exterior form.

Objection. They say the Virtue of Elements is their Materials.

Answer. Right! In respect of their beginning upon pure Bodies.

Objection. All their Secrets spring from One Vile thing, common to Rich and Poor.

Answer. Precious things corrupted are most vile, and Science is common to Rich and Poor, and have much use of Calx or Dust.

Objection. No Metal is required to the making of the Stone.

Answer. As no part of Man to the making of Man.

Objection. One thing, one Glass. One furnace is sufficient.

Answer. True, where two things of one kind be apted and conjoined.

Objection. Out of one Root proceeds White and Red.

Answer. Even as Male and Female from one Womb.

Objection. The Stone is Vegetable, Animal and Mineral.

Answer. Right! Joint and several, it is said to be vegetable because in the Maturation, it is multiplyed in Virtue and quality. It is said to be Animal because it increaseth his own kind. And it is said to be Mineral because his Original is from the

Metal, or their Mineral. Here we may remember the Bishop of Utrecht, who lost his Life for discovering his Secret. Why should we prevent the highest distribution, who hath not made knowledge hereditory, but wrapped things in secret, that we might difference things in being, and in being and use. Nature is ever jealous of her supremacy, and abhoreth to see the sensible before the Intellectual treasures preferred. This shews the beginning and the end of Art. Lux sata est justo eum rectis animo lactitia. Mark what ye sow, for such is your Harvest. Light is sown on pure Earth; Some grain begins to put forth Ear at three joints, some at four, but the Ear never buds until the joints be grown. And what Virtue this knotting or fixing gives, consider, for by meditation you may see, by seeing you may know, by knowing ye delight, by delighting ye adhere, by adhering ye possess, by possessing ye enjoy the Truth, that is the uncorrupted Nature made visible, therefore take heed how you value.

It is held, a long strife is a greater hurt, for the clearer sight ought to direct, but we must leave the world to Opinion, yet because Truth fears nothing but to be hid, and desires nothing save not to be condemned before she be understood. Our Love to all Truth, shews an awful assertion to the sovereign Truth, which is not subject to ocular demonstration, because Truth and Religion are Universal, and the Sense only capable of particulars, and an Art done without knowledge of the Cause, is but a fallacy.

It is observed the Protestant by the purity of his Doctrine infers the verity of the Church; but the popist would prove the purity of his Doctrine by the visibility of his Church, and so prefers personal confidence before rational demonstration. The like difference is found amongst the Chemists, the one proves the verity of his Art from the nature of Invention, showing the Effect from the Cause, by the nature and property of the Subject, even unto the third and fourth kind of qualities, for Science and Essence are one, as an honest Mans Word is a deed, yet the End is concealed, because it is of infinite virtue. The other would prove the verity of his Art by Blanchers, Cementations, and Citrinations, which only seem, but are short of perfection. The true Elixirs are exampled by Urim and Thummim, which are joint and several and convertible terms, the names signifying the thing, and the Thing predicating the Name, for they signify Light and Perfection and were Corporate and perfect Lights.

The Rabbins affirm in the Second Temple they made Urim and Thummim, but did not enquire of God by them, because Nature and Art is not sufficient unto Gods service, without His own sacred Ordinance.

Moses had a Command, and did put them in the double fold of the Breast-plate. Therefore they were Substances, and none but the King or General might enquire, as the Philosophic precept is, not to speak of God without knowledge. They are numbered amongst the Artificial Things (Gen. 3:4). And Bezaliel the Son of Uri, which signifies Light, was commanded to devise works upon SOL, which is called the Great Secret of the Almighty.

The Philosophers remote Beginnings for their Elixirs are the same, the one is called Donum Dei, so is the other; And the Magnalia Dei. Their extractions of the Central Virtue; Their Operations and Progressions are alike, therefore necessarily the effect must be of like infinite Virtue, and next to the Rational Soul, the richest Treasure. How pitiful a presumption is it to mount the Chariot or climb Hermes Tree without a Golden Ladder, few are understandingly able to discern a true good, but it is Textual advise, to buy ye Truth (Prov. 3:14) although the Justice of the price doth not always pitch upon a point. Each thing in this Art is wonderful, and as in true Marriage virtues are matched, so in the Marriage of the Elements the Visible things are of small continuance, although from the end of every intention springs the beginning, and pure beginnings are from the Virtue of the Elements which are Eternally fixed.

If you observe how Heaven and Earth are disponsate you shall find nothing hid from the heat of the Celestial Sun, and the form of the Terrestrial Sol is Celestial Virtue. So innate and infused Virtue drawn from the Centre are Eternally fixed, most durable, and of infinite Virtue, but things of greatest increase are of longest expectation, and the most pure most suffering. If Knowledge have begot affaction you shall think the time short, because of the excellency of it, for consolation follows patience, and thereunto only is Victory promised. When you see a Linear course in the parallalism of the Regeneration of Man, and the Purification of Metal, why doth tract of time daunt, when Art doth that in short time which Nature doth in one thousand years.

The making of Urim and Thummim and the perfecting of the Elixirs is aptly compared to the fourfold creation of Mankind. Adam from the Earth, Evah from Adam, Abel from both, and Jesus Christ from a Virgin. So Man called a Living Stone, produceth that Eternal Stony and Fire conquering Spirit, called the Elixir, from their proper Earth, only their Adam, from him their Evah, from both their Virgin, from her only the Sovereign and Universal Spirit, which doth vivify and preserve all loving Creatures, and raiseth the Artist from the dust, to sit amongst Princes.

Life without Sin is Wisdom-manifest in the Flesh.

A Body without Shadow is the Universal Spirit corporate.

F I N I S.

AN HUNDRED APHORISMS

CONTAINING THE WHOLE

BODY OF MAGICK

anonymous

1. The whole World is animated with the first supreme and intellectual Soul possessing in itself the seminary reasons of all Things, which proceeding from the brightness of the Ideas of the first Intellect are as it were the Instrument by which this great Body is governed and are the Links of the golden chain of Providence.

2. While the Operations of the Soul are terminated or bounded the Body is generated or produced out of the power of the Soul, and is diversely formed according to the imagination thereof, hence it hath the denominating power over the Body which it could not have except the Body did fully and wholly depend upon it.

3. In this production whilst the Soul fashioneth to itself a Body. There is some third thing the mean

between them both by which the Soul is now inwardly joined to the Body, and by which the Operation of all natural things are dispensed, and this is called the VITAL SPIRIT.

4. The Operations of Natural things are dispensed from this Spirit by the organs according to the disposition of the Organ.

5. The disposition of the organ depends first and primarily upon the Intellect, which disposeth all things. Secondly upon the Soul of the World that forms itself a Body according to the seminary reasons of things. Thirdly, unto the Spirit of the Universe that containeth things in such a disposition.

6. No bodily thing hath any energy or operation in itself saving so far for as it is an Instrument of the Same Spirit, or informed by it, for that which is meerly corporeal is meerly passive.

7. He that will work great things, must (as much as possible) take away corporeity from things, or else he must add Spirit to the Body, or else awaken the sleeping Spirit, unless he do some of those things or know how to join his imagination to the imagination of the Soul of the World, now labouring

and undertaking an exchange, he will never do any great matter.

8. It is impossible to take all this Spirit from anything whatsoever for by this Bond a thing is held from falling back into its first matter or nothing.

9. This Spirit is somewhere or rather everywhere found as it were free from the Body, and he that knows how to join it with a Body agreeably, possesseth a Treasure better than all the riches of the world.

10. The Spirit is separated from the Body as much as it is possible either by means of fermentation or drawn by his Brother which is at liberty.

11. The Organs by which this Spirit worketh are the qualities of things, which meerly and purely considered are able to do no more than the eyes can see without life, as being nothing else but modification of the matter of the Body.

12. All things operating do it to this only purpose, to make things upon which they work like themselves.

13. The Subject of the Vital Spirit in the Body in it is received, and by it worketh, neither is it

ever so pure but that it is joined with its Mercurial moisture.

14. This Humor doth not specify the Spirit because it is the common matter of things apt to be made anything neither is it seen with eyes because it is pure, until it be first terminated in a more solid body.

15. Neither Souls nor pure Spirits, nor Intelligences can work upon Bodies but by means of this Spirit, for two extremes cannot be joined without a mean, therefore Demons appear not but after sacrifices used.

16. If the Spirit or Intelligencer of it be specified with being, either dissipated by the contrary or changed into another thing, they cease to work there any longer, as they are allured by the Vital Spirits of living creatures so they are put to flight, or rather cease to work upon Bodies, where sharp and venomous things are used.

17. The Stars do tie the Vital Spirit to the Body disposed, by light and heat, and by the same means do they inspire it into the Body.

18. In Generation the Spirit is mixed with the Body, and directs the intent of Nature to its end.

19. The Seeds of things are known to contain more plenty of this Spirit than anything else.

20. The Seeds do not contain such plenty of this Spirit as is requisite to the perfect production of a thing, but the internal Spirit allureth the External coming down from Heaven unites it to itself, and being fortified therewith at length it begets its like.

21. Before the seeds do germinate or bud, it is fermented, and by fermentation disposed to attraction.

22. If germination should be hindered with the advancement of attraction and assimilation, the things might be brought at length from the seed to the spirit of it in a moment.

23. That which is more universal doth more further attraction and more disposeth their seeds to attraction as Salt-petre in vegetables.

24. Every family of things hath some with its mixt universal, whereby the seeds are disposed to attraction and made fruitful.

25. He that knows how to join artificially the Universal to the seed of the Animal family may produce everlasting Weights, besides the terminate matrix or womb at least formally, and the like reason 'tis also of other things.

26. He that can join Light with Darkness can multiply things in their own kind, and change the nature of them.

27. The Universal Vital Spirit coming down from Heaven, pure, clear, and uncontaminated is the Father of the particular Vital Spirit which is in everything, for it increases and multiplies it in the Body; from whence the Bodies borrow the power of multiplying themselves.

28. As the first Vital Spirit lies in the Mercurial Humor that is common and free, and the vital Spirit of particular things is resident in that Mercurial Humor imbued with the virtue of that Body whose it is, which they call radical Moisture.

29. He that can join a Spirit impregnate with the virtue of one Body with another, that is now disposed to change, may produce many miracles and monsters.

30. The First Variety of the Disposition of Bodies proceeds from the Various concoction of Water.

31. The second, from the various mixtures of the three principals, Salt, Sulphur, and Mercury.

32. These dispositions flow from the various positions of the Stars, especially from the Sun.

33. Everything hath so much Vitality as is required to produce the natural actions of the species.

34. Nothing begins to be made that doth not receive some Vitality from Heaven by which it can work somewhat.

35. He that knows how to infuse the propitious Heaven or Sun into things, or the mixtures of things, may perform wonders, and hereupon depend all Magical Operations.

36. By how much the Disposition or the Subjects are more formal, so much more of this Life they receive, and so much more powerfully they work.

37. As in the eye, the operations are more noble than in the foot, although they both proceed from the same Soul, because of the variety of this Organ apt to receive a greater protion of Life; So the constellate characters because of their formality receive a greater portion of Spirit from Heaven and perform noble actions.

38. This Spirit continually flows from Heaven and back again to Heaven, and in the flowing is found pure and unmixed, and therefore may by a skillfull workman by wonderful means he joined to anything, that increase the virtues of it according to the disposition of the subject.

39. The Heart of Heaven is the Sun, which by Light distributes all things, as well to the Stars, as to the Earth.

40. Opaque is nothing else but a Body either wanting Light or having the Light asleep in it.

41. He that can by Light draw Light out of things, or multiply Light Sith Light, he knows how to add

the Universal Spirit of Life to the particular
Spirit of Life, and by this addition do miracles.

42. How much Light is added, so much Life, and so
much of the one as is lost, so much is lost of the
other.

43. This Spirit, after the last process of
Maturation, straight begins by little and little to
vanish.

44. Maturation is nothing else but the operation of
the proper radiated Spirit to the perfection of the
Individual, as far forth as it may be perfected,
proceeding to the seminary reasons expounded or
proposed by Nature or the Soul, or it is an
actuation of the Internal Spirit as far as it can be
actuated, or it is the greatest Illumination of the
matter that can possibly be done by such Light.

45. The Spirit is dissipated when it strives to act
upon a matter too rebellious, or when the natural
Crassis or mixture of a thing is altered by the
Stars, sometimes too much excited it breaks forth,
or being called out by his Brothers Spirit it goeth
away to it.

46. The Matter is rebellious when by reason of a Starry Crasis or temperature it cannot be overcome allured by the Spirit, or when it is in the natural periods beyond which it can neither go, nor the Spirit carry it any further, for only so much Spirit is given as serveth everything to the due perfection of it.

47. The Temperature of a thing is altered by the Stars which the Horoscope of the Nativity comes to the degrees or positions of the Planets contrary to the beginning of Life.

48. The Spirit is too much excited by fermentation or immoderate agitation, for, moderate agitation is necessary to vital operations.

49. The Spirit is called out by his Brother Spirit when it is too much exposed to it.

50. In certain things it cannot be called out by its Brother, because of its straight locality with the Body, but it allures its Brother to him and is strongly fortified thereby.

51. Fermentation is the action of heat upon moisture, by which the moisture is treated and made subject to the Spirit, or it is the affect of the

Spirit circulating itself in the Body which cannot remain in the same state because of the fluxibility of the Body.

52. He that by means and use of Universal Spirit can excite the particular Spirit of anything to a natural fermentation, and appease and settle the natural tumults by repeating the operation, may miraculously increase things in virtue and power, the highest Secret of the Philosophers.

53. Every man knows that by means of fermentation the spirit is as pure as it possibly may be drawn, but almost all men do it without the fruit of multiplication, because they know not how to join one Brother with another.

54. Things do abide in the same state of nature so long as they possess so much Spirit as is sufficient to perform the due execution thereof.

55. Everything fermented works more strongly because in things fermented the Spirits are more free.

56. Hence is manifest the cause of the natural Death or destruction of things, everything tends to maturation as to the perfection thereof, and when its (?) the Spirit begins to show its force, and so

by acting it is dissipated and vanisheth, which at length is the cause of destruction.

57. He that could lay hold on this vanishing Spirit and apply it to the Body from whence it slipt, or to another of the same Spirit, may thereby do wonders.

58. From this fountain are all natural Philtres flowed; for easily may the Spirit be imbued with the qualities of another Body causing in Bodies of the same kind a real similitude, which is the violent cause of Love.

59. These things are apt to intercept this particular Spirit which have the greater similitude of most natural conjunction with the parts, or which being applied to a vegetous body, are by such a contact made more flourishing. These things are to be understood of the Bodies of Wights, especially of Man where Philtres are of most power.

60. This Spirit where it findeth a little Matter disposed according to that likeness, it makes and seals the compounds produced.

61. Where the Spirit of one Body being married to the qualities of that Body is communicated to another Body, there is generated at a certain

compassion, because of the mutual flux and reflux of the Spirits to their proper Body which compassion or Sympathy is not easily dissolved as that which is done by imagination.

62. There can neither Love nor compassion be generated without the mixture of Spirits.

63. This commixture is sometimes done by material application, sometimes by imagination, and not seldom by the disposition of the Stars.

64. By natural application it is done when the Spirit of one Body is implanted in another, by means of those things which are apt to intercept the Spirit, and to communicate it to another, and they are known by the signature, and by the Ancients called Amatoria or such things as love one another.

65. By imagination Love is produced when the exalted imagination of one doth predominate over the imagination of the other, and so fashioneth and sealeth it, and this may be easily done because of the volubility of the imagination. Hence all incantations get their efficacy for though peradventure they have some vertues in themselves, yet this virtue cannot be distributed because of the universality thereof.

66. From the Stars Love takes its beginning either
when the disposition of the Heavens is alike at the
times of Nativity as the Astrologers do abundantly
teach, and this is most firm and most to be desired,
or when the beneficial Beams of the Stars being apt
for the purpose are at a fit time received into
matter disposed and in a due manner brought into
Art, as Natural Magick more fully teacheth.

67. He that can to these manners of doing add the
Universal Spirit may do wonders.

68. Thou mayest call the Universal Spirit to thy
help if thou use instruments impregnated with this
Spirit, the greatest secret of Magicians.

69. He that knows how to make a vital particular
Spirit, may cure the particular Body whose Spirit
that is at any distance, always imploring the help
of the Universal Spirit.

70. He that can fortify the particular Spirit with
the Universal may very long prolong his Life; unless
if Stars be against it, yet by this means he may
lengthen his Life and Health, and some state the
malace of the Stars as he needs confess that knows
the habitation of this Spirit.

71. Nothing can be putrified without it first feel fermentation, because nothing comes naturally to inclination but by state.

72. Putrification is the symptom of declining nature, or of the spirit flying away.

73. There is nothing putrified that hath not great store of the volatile spirit.

74. All heat proceedeth from the Vital Spirit, and is said of motion neither can that Spirit subsist without heat, or at least not be mingled with bodies.

75. Everything that is putrified hath less heat in it than it had before the putrifaction, therefore it is false that things putrifying do grow whole.

76. As much spirit so much heat is gotten, and of the one is lost so much as of the other.

77. Heat can neither be stirred up by nature nor Art, but by means of Light, either external or internal.

78. He that shall call Light the Spirit of the Universal shall peradventure not far miss the truth, for it is either Light, or hath his dwelling or habitation in the Light.

79. He that can destroy bodies without putrifaction, and in the destruction can join Spirit with Spirit by means of heat, possesseth the principal Secret of Natural Magic.

80. The external Light heateth by bringing in a new heat, and by actuating its own heat, whether it (the Light) be determinate or indeterminate.

81. THE LIGHT DETERMINATE POSSESSETH A DESTROYING HEAT, and such an one as burneth all things, so it be compactly actuated as in fire.

82. Indeterminate Light giveth Light, and never hurteth anything but by accident.

83. He that knows how to make Light determinate of Light indeterminate, not changing the Spirit, nor receiving it otherwise than in a common medium knows exceeding well how to purge mineral and all hard bodies without loss of radical moisture.

84. The Light which we call determinate and which hath in it the Life of things being the carriage of the Universal Soul lieth hid in darkness, neither is it seen but by a philosopher to whom the Centre of Things is apparently discovered.

85. The internal heat is excited by reason of the internal Spirit whose house it is.

86. The Spirit is agitated by fermentation or motion, sometimes they occur or concur both together to agitation.

87. There is a third secret means of Agitation known to the Philosophers which is perceived by them in generation and regeneration.

88. When fermentation is distinguished from motion understand local progressive motion which comes from the imagination directing the vital Spirits to motion.

89. All fermentation finished before the due time is a sign of moderate putrefaction succeeding.

90. He that knows how to hasten fermentation and hinders putrefaction by having the Spirit of the

Universe propitious, understands the Philosophers contrition, and can by means thereof do minerals.

91. Putrefaction hath not its original from the Spirit, but from the Body, and therefore it was contrary to the Spirit.

92. He that knoweth the Spirits of the Universe and the use thereof may hinder all corruption, and give the particular Spirit the dominion over the Body; how much this would avail to the cure of diseases let physicians consider.

93. That there may a universal medicine be given is now agreed on all hands, because if the particular Spirit get strength it can of itself cure all diseases, as is known by common experience, for there is no disease which hath not been cured without the Physcans help.

94. The Universal Medicine is nothing else but the Vital Spirit multiplied upon a due subject.

95. He that seeks this Medicine elsewhere than in the tops of the highest mountains shall find nothing but sorrow and loss for the reward of his pains.

96. The Philosophers who say it is to be sought in the Caverns of the Earth mean the Earth of the Living.

97. They that hope to find it in the furnaces of the Chemics are desperately deceived, for they know not the fire.

98. Nothing hath from the first intention of Nature more Spirit than is sufficient for it, to the conservation of its proper Spirit, yet out of everything Nature playing the midwife for him, the philosopher can produce a Son nobler than his Father.

99. The first and the last colour of things are yellow, because the Stars and the Sun are Yellow, those things that are of a lesser temper as the planets appear Green, after they have touched the Air, being naturally and most highly ceruleous or Blue, and working upon them makes Yellow things Green, but being made harder they put on again their first and natural colour, out of the things that have been said thou mayest pick up Mysteries.

100. The Air is Blue, and the Horizon appears Blue to the sight on a clear day, and the Air by reason of its thinness is not apt to terminate the strong

and rigerous Vital Beams, until they languish and
grow weak by distance, but then the terminated Beams
show the native colour of the Air, And thus much to
have said at this time by way of Aphorisms, if thou
make not very much accompt of it, is too much.

Finis.

THE GREAT WORK OF THE LAPIS SOPHORIUM

ACCORDING TO THE LAMSPRING pnocess

OF THE RUBIFICATION OF THE WHITE

SULPHUR NATURAE EXMERCURIUS

Having obtained the White Sulphur of Nature from Mercury in two or three digesting globe glasses, take that glass which you propose to continue to digest till it be perfected into the Red Sulphur and without permitting it to cool place it in a lamp furnace in a bed of sifted ashes, warmed to the same degree of heat as the glasses had acquired in the water bath. The dry heat in ashes must be no stronger than that you can bear the glass in your open hand. Continue the gentle degree of dry heat, say about 120 to 130 degrees until your SULPHUR NATURAE ALBUM is become of a very bright and beautiful cinnabar colour, which it will in about thirty days; This is Sulphur Rubrum Naturae Indeterminatum.

Solution of the Red Sulphur

Naturae into an Oil.

Dissolve this Red Sulphur of Nature by the same
process as you did the white Sulphur; that is
dissolve it in a genuine highly rectified Spirit of
Wine, digest in a blood warm water bath, keeping the
glass close shut and you will obtain a deep ruby red
transparent solution. This solution is FIRE.

If you tinge a bottle of good old White Rhine wine
or Austrian wine with this essence, until the same
becomes as deep in colour as Burgundy, which a small
quantity of the dissolved red sulphur will effect,
you have then in your possession. The Glorious
inward Medicine; or Quinta Essentia Medicinalis
which is so powerful that a few doses of a coffee
spoonful will expel the most dreadful diseases,
Epilepsy, palsy, dropsie, consumptions, fevers,
Gout, leprosy, all fly before it.

It is a cure for the maladies of the whole animal
creation.

But when the Solar Sulphur spiritualized, has been
united and coagulated therewith, it then becomes a
hundred times more powerful, and must therefore be

dilated proportionally before it be exhibited as a medicine. One single grain in substance in that state would extinguish life like a stroke of lightning or a Violent shock of electricity which is the same thing with less power, as we have proved by experiments made on dogs and other animals.

Distillation of the Ruby Red
Transparent Solution of Red Sulphur of Nature.

Having, by the means directed, obtained your ruby red transparent Solution of the Red Sulphur of Nature in Spirit of Wine, you must with a gentle heat in Balneo draw off the Spirit of wine, per Alembicum until there remains behind a ruby red oil.

Composition of the Principles.

To three parts of the Ruby coloured oil you must add one part of the GOLDEN FERMENT reduced to an oil, by means of spirit of wine.

Manage exactly as you did the White, and Coagulate the united oils in a digesting globe glass placed in a dry heat of sifted Ashes, leaving the glass open during the first 24 hours of digestion to evaporate the superfluous humidity. Then shut it and digest until it is become a beautiful DEEP RED MASS. This will be soon accomplished in a heat of from 120 to 130 degrees. The trial is that it must melt without fuming.

Multiplication in quality, virtue & Power.

The Multiplication of the red is performed in exactly the same manner as that of the white tincture formerly taught.

You must dissolve the above red mass (which is the RED TINCTURE IN AN INFANT STATE capable of transmuting ten parts only of Mercury into SOL) in your rectified lac virginis by a gentle digestion.

When perfectly dissolved distill the Mercurial Spirit from the tincture until it remains an oil.

This being put to a digesting globe, placed in warm ashes, must be dried up again until it become again A RED MASS.

Repeat this solution and coagulation, until it will not dry up any more, but remains a FIXED RUBY RED OIL which shines in the dark. This is the Elixir

Reberum tertial ordinis, which is capable of
vitrifying a great quantity, at least one hundred
parts of refined gold in a crucible, which vitrified
gold can convert a greater quantity, at least 1000
parts of Mercury into a RED TINGING CINNABAR, or
precipitate, which, finally can transmute at least a
hundred parts of Mercury into fine gold.

The red tincture is capable of being still further
multiplied.

Before it has vitrified gold, it is the lapis
Sophorum MEDICINALIS UNIVERSALIS, the URIM AND
THUMIM which gives light in the dark and tinges
alcohol of wine into a ruby red essence, wherewith
you can tinge old white Austrian wine into the
medicine, capable of healing and overcoming all
diseases, and able to preserve life beyond the
general term.

The dose of this tinged Wine must be small, a few
drops only and that not too often.

To Prepare the Solar Ferment.

Take the pure gold of 24 Carats refined with the
greatest care by a faithful refiner, two ounces; get

this heat into thin leaves at a gold beaters, one whom you can trust and who will not change your gold, you ought to get enough beat to yield you two ounces of leaves.

Dissolve the gold leaf, one leaf after another gradually in your lac virginis; mixed with a a of good aqua fortis in which aqua fortis you have previously dissolved one fourth part of its weight of Sublimed Sal Ammoniac to make it become an aqua regia.

Let your double solvent consisting of your aqua regia just mentioned and your lac Virginie, of each an equal weight, weigh twice as much as your gold does, that is, have four ounces of solvent.

Dissolve the gold leaves gradually, without heat and you will obtain a beautiful transparent fiery red liquor.

This is the humid calcination; shut the digesting glass and place it in a blood warm water bath to digest for eight days.

Then distill the solvent from it very carefully until there remains behind an oily liquid gold.

Digestion.

Put the SOLAR OIL just obtained into a digesting globe glass and set it in a water bath of a blood heat for one hundred and fifty days (five months) and the gold will die & rot as the silver did before.

After blackness is over you will obtain in about six months time the White Mercurial Sulphur of gold which will settle all round the globe like small pearls or the eyes of fish.

Rubification of the White Sulphur of Gold.

When you have the sign just mentioned, your White Sulphur of Gold settled round the globe like small pearls, take your glass gently out of the water bath and place it in ashes previously warmed over a lamp as nearly the same degree of heat as the water bath was, then increase your heat gradually to 110, 120 and 130 degrees and the White Sulphur will change into a yellow and finally into a beautiful deep red colour.

The change from the white to the red will be accomplished in five or six months (weeks) and you

will then have in your possession the red
spiritualized gold or Solar Ferment extremely
fusible.

Solution of the Solar Ferment and reduction of the same into Oleum SOLIS (Solaris).

Dissolve your red solar ferment in genuine highly
rectified alcohol of wine and you will have a
Transparent ruby coloured solution which no art can
reduce per se into SOL again.

This ruby tincture is AURUM POTIBLE PER SE but not
LAPIS PHILOSOPHORUM MEDICINALIS. Yet it is a
glorious restorative and curative Medicine.

Distill the spirit of wine in balneo gently from the
solution, per alembicum, until there remains behind
in your glass body a deep ruby red oil of gold, that
is a solar oily looking liquid, which is the
Spiritual Solar Ferment for the composition of the
Red Elixir primae Secundae et tertiae ordinis.

Soli Deo Gloria.

Finis.

LA LUMIERE DU CHAOS

par

LOUIS GRASSOT

Amsterdam: 1784

Translated from the French

by

J.W. HAMILTON-JONES

London: 1953

LIGHT EXTRACTED FROM CHAOS

PREFACE

Philosophy took birth with the beginning of the World. At all times men have thought, reflected and meditated to find ways to live as a community, but self—preservation is not without interest and we may well think men forgot about this being so much occupied by their surroundings; subject to so many vicissitudes, the butt of so much that is bad, men seek to enjoy those things which surround them. Without doubt they have sought means to prevent illness and also the remedies which will restore them to good health and preserve it for as long as possible to which end they lend a willing hand in order to escape disease. Neither have men failed to reason upon the Beings of the Universe and to meditate deeply to discover that fruit of life and that source of riches capable of bringing man near to immortality; that they are not mistaken in this is supported by the fact that in our days there exists a man named M. de St. Germain, one of the most famous adapts of the century, who with this precious treasure which he possesses has attained the age of more than four centuries, and is still alive, free from all those infirmities usually brought on by old age, and enjoying a fortune to his taste. Secondly, it was announced in the Journal Encyclopedique de Bouillon on 1st. February 1783, on the subject of the transmutation of metals that in England by means of the powder of projection this work had been demonstrated. Neither can we doubt the fact, it having been effected in the presence of magistrates and other responsible witnesses above suspicion, who affirm the truth of this experiment.

Practical demonstration of a treasure of this nature
is not novel, but it is nevertheless usually
confined within a very small circle of people, who
think that because God has not given this knowledge
freely to all men, He does not wish it to be
divulged, and therefore those who possess the secret
make it known but to very few selected friends.
Hermes Trismegistus, or the "Thrice Greatest", the
first amongst all the Philosophers known to be
distinguished, would not communicate this work
except to the elite, after having proved them to
possess both prudence and discretion, and those
passed it on to others who were worthy, and of
similiar qualifications.

How is it possible to communicate from age to age
these admirable secrets, and at the same time to
conceal them from the public eye? If it were done
through an oral tradition there would be the risk
that it would be lost for future use, memory being
too feeble a thing in which to place complete trust,
and traditions of this kind become obscured by the
passage of time; the further they get from their
source the more impossible it becomes to fathom the
dark chaos in which they are wrapped. There is no
other way but to have recourse to hieroglyphs,
symbols, allegories, fables and similar methods
which are susceptible of many different

explanations, which serve to inspire changes in interpretation, and thus to instruct some people whilst others remain in complete ignorance. This is the method chosen by Hermes and after him all the Hermetic philosophers have done the same, causing amusement to the people by fables, as says Origen, and these fables together with the names of the Gods used in the country serve to veil their philosophy. It is now time for this Veil to be rent so that light may appear from chaos, and be shown in all its brilliance; Hippocrates must break his silence because I regard it like a theft which man commits against society, to conceal the discoveries which he has been able to make which would promote happiness and a general preservation from maladies.

I hope that those who apply themselves to this science will appreciate the trouble I have taken to compile this small book in the most intelligible manner possible.

The whole operation of the grand philosophical work has been made difficult of access and wrapped up in allegories. Maybe I shall not merit the approbation of those great subtle and penetrating minds whose knowledge embraces all things; who know all without ever having learnt anything; who discuss every subject and arrive at a conclusion without knowing the cause; therefore it is not to such people that one gives lessons because to them properly belongs the name of Sage, even more so than to Democritus, Plato. Pythagoras and other Greeks who went to Egypt to breathe the Hermetic air and draw on the science of which this book treats. When one needs light on a subject which is difficult to believe for the sole reason that it is rare and extraordinary it is prudent to remember this verse from Lucroce:—

"Although reasoning cannot discover the cause, it is true."

The first man to conceive the idea of flying in the air, was ridiculed as soon as he mentioned it and treated as a senseless fool, but this did not prevent many other people from seeking the means and not being discouraged, although they were told it was an impossibility. Now, in our days, we see with great satisfaction that M. de Montgolfier has found success in his enterprise, which proves that everything which is conceived by the spirit in man is possible, and that it all depends on finding the means of arriving at our object and working on true principles.

If the incredulous and prejudiced will take the trouble to follow step by step the route which I shall mark out, they will see, to their great astonishment, the true banishment of the spirit of unbelief and fear, which may have been occasioned by their experience with a number of puffers and charcoal burners who do not succeed in their experiments because they work on false theories and do not know the First Matter without which one can do nothing and should never undertake any enterprise because this knowledge is the fundamental and general basis of the Philosophical work.

Finally, I beg the reader to be persuaded that I have no other interest in view than to demonstrate the Truth to those who aspire to know it, and I desire, with all my heart, that those who unhappily lose their time in working with substances which are foreign and out of line, will receive enlightenment by reading this book and come to know the true and unique subject of the philosophers; and that those who already know it but are ignorant of the great point of the dissolution of the stone, and the coagulation of the water and the spirit of the body which is the completion of the universal medicine, will be able here to apprehend the secret operations which are described so accurately.

LIGHT out of CHAOS

or

THE HERMETIC SCIENCE

Of the Great Work of Philosophy

Which out Ancient Sages produced

The Source of Riches and of Good Health

THE KEY TO NATURE

Out of every material thing there is produced a cinder, from the cinder there is made a salt, from the salt there are separated water and the mercury, out of the mercury there is produced or prepared an Elixir or a Quintessence. The body is reduced to cinder so that it may be cleansed of its combustible parts; to salt to separate it from its earthliness; to water to purify and putrify; to spirit so that it may become Quintessence.

Thus the salts are the keys both of art and of nature; unless they are known it is impossible to imitate nature in her operations; it is necessary to know their sympathy and their antipathy towards metals as well as amongst themselves; properly there is only one Salt of Nature, but it divides into three sorts to form the principles of bodies; these three are Nitre, Tartar and Vitriol, all other salts being composed of them.

Sublimation, precipitation and coction are three methods which Nature uses in the perfection of her works; by the first she takes off the superfluous humidity which would suffocate the fire and prevent its action in the earth, which is its Matrix.

By precipitation she returns the humidity to the earth of which the vegetable kingdom and the heat have deprived it. Sublimation is performed by the elevation of the vapours into the air where they are condensed into clouds. The second method is done by the rain, with rain and fine weather alternating; a continuous rain would drown everything, perpetual fine weather would dry up everything. Rain falls in drops because if it were precipitated in volume all would be destroyed; no gardener water his seed from a full jet; it is thus that Nature works and distributes her blessings with weight, measure and proportion.

Coction is a digestion of the crude humour instilled in the bosom of the earth, a maturation and a conversion of this humour into aliment through the medium of her secret fire. These three operations are so linked together that the end of one is the beginning of the other.

Sublimation has for its object the conversion of a heavy thing into a light one, and exhalation in vapours, to attenuate a gross and impure body, to cast off its faeces, to make the vapours take up the virtues and properties of superior things, and, in effect, to relieve the earth of a superfluous humour which would impede her productions.

No sooner are these vapours sublimed, than they are condensed into rain, and, spiritual and invisible though they be, they immediately become a dense and aqueous body and fall upon the earth to imbibe it with that celestial nectar with which they have become impregnated during their sojourn in the air. As soon as the earth receives them, Nature works to digest and to cook them.

The water contains a ferment, a vivifying spirit which trickles from the superior parts upon the inferior by which the superior were impregnated during their wandering in the air, and which they now in turn deposit in the bosom of the earth. This ferment is the seed of life without which men, animals and vegetables would neither live nor

multiply; everything in Nature breathes this thing
and man does not live by bread alone but by this
aerial spirit which he absorbs ceaselessly.

Only God and His minister Nature, know how to bring
the material elements, the principles of bodies,
into obedience; art does not know how to attain this
without the three which become tangible as a result
of the resolution of mixed bodies, Chemists name
these Sulphur, Salt and Mercury; these are the
principles and elements; Mercury is formed by a
mixture of water and earth; Sulphur of earth and
air; Salt of air and water condensed together.

The fire of Nature is joined as a formal principle;
Mercury is composed of a fat viscous earth and a
limpid water; Sulphur of an earth very dry and very
subtile mixed with humidity of the air; Salt, in
fact, of gross pontic water and a raw air which is
found enclosed therein.

Here is the explanation of subterranean physics by
Becher, upon this subject.

Nature is very simple in her operations, therefore
it is necessary to imitate her if we would be
successful in the enterprise; she has but one
principle, and also there is but one fixed spirit,

composed of a very pure incombustible fire which remains as a radical humidity in all mixed bodies; it is more perfect in gold than in any other substance, and only the philosophers' mercury has the property and virtue to extract it from its prison, to corrupt it and to render it fit for generation; quicksilver is the principle of volatility, of malleability and of minerality; the fixed spirit in gold can do nothing about this; gold is humected, reincrudated, and volatilized and brought to putrefaction by the operation of mercury, and this is digested, cooked, thickened, dried and fixed by the operation of the philosophical gold which renders it, through these means, into a metallic tincture.

One is the philosophical mercury the other is the sulphur; this sulphur is the soul of all bodies and the principle in the extraction of their tincture, but common mercury is deprived of it and gold and silver vulgar have but sufficient for themselves. The mercury appropriate to the work must, therefore, first be impregnated by an invisible sulphur so that it will be more disposed to receive the visible tincture of the perfect bodies and afterwards be able to communicate it with profit.

Many chemists sweat blood and water to extract a tincture from common gold; they imagine that by the force of the torture they give it, they will discover the secret of its augmentation and multiplication but:

"The hungry plowman is cheated by vain hopes", because it is impossible that the solar tincture can be entirely separated from its body; art does not know how to destroy in this metallick genus that which Nature has united so well; and if they could succeed in extracting from gold a liquor both coloured and permanent, by the force of fire or by the corrosion of strong waters, the result would be regarded as a part of the body and not as its tincture because that which properly constitutes the tincture, cannot be separated from the gold.

OF THE PRIMARY SUBSTANCE

which is the only substance to be used to make

THE POWDER OF PROJECTION

The source of health and riches, twin bases upon which the happiness of this life is sustained, are the objects of this art which has always been a mystery; and those who treat of it, have at all times spoken of it as a science, the practice of which has something of the supernatural about it, for its results are miraculous both in themselves and in their effects.

In spite of all the information which one may give conducive towards the knowledge of the primary substance, the Great Architect of the Universe, Creator of all Nature, Whom the philosophers propose to imitate, alone can illumine and guide the human soul in the search for this inestimable treasure, as well as in the operation of this art.

Therefore, if you desire to succeed, seek in His name and you will find a substance which is the daughter of the sun and of the moon, which contains within itself the four Elements as well as the three Kingdoms of Nature through which everything exists. This matter has no fixed or determined shape except that it is flat, green, membranous, gelatinous, without root or branches; in fact, its shape and the manner in which it is produced as well as its essence, have made men give it the names of sperm of the earth; Heavenly blossom or Nostoc; in effect she resembles a green sperm which is spread over the earth in particles or fragments of unequal size. She is found in the uncultivated parts of the earth which are slightly moist and mossy and abound in long, narrow, stony and sandy pathways, near to the mountains; in fact, she is to be found everywhere. She must be gathered before sunrise in the spring after the 21st. of March until the 21st. of April, and in the autumn after the 21st. of September until the 21st. of October. That which one gathers in the springtime is the female, and that of the autumn is the male; it is desirous to gather the greenest. Understand that you will put into work the quantity you have gathered in each season. I must tell you that the essence of this substance is held in the air with the celestial body, having both masculine and feminine qualities, of firm and strong virtue, fixed and permanent and that it is carried by the air into the bosom of the earth which serves it as a matrix, thence to corporify it; therefore the sun and the moon produce it from their fecundity; which circumstance has caused the Hermetic Philosophers to give it the name of Son of the Sun and of the Moon, this name belongs to it more properly than all the others, and it has been given to hide it and conceal it from the eyes of the vulgar. It is necessary, therefore, before one is able to understand anything, to know this matter, the pure and the impure, the clean and the foul, because nothing in Nature can give that which it does not possess: and for this reason things are not, and cannot be,

different from their nature or from their
principles.

Take therefore, the part which is nearest and which
is most perfect, and it will suffice; leave the
mixed and take only the simple, because it is there
where one finds the quintessence and by these means
you will make the medicine which some people will
call quintessence which is the principle which
cannot perish, is permanent and always triumphant.
It is a brilliant light which truly illuminates
every soul who has come to know it; here is the knot
and the bond of all the elements, which contains in
itself the spirit which nourishes all things and the
means whereby Nature is stirred throughout the
entire universe; this sprouting fountain is the
commencement and the end of all her operations.

I counsel you therefore to reject every other thing
as useless and to take only this water which burns,
whitens, dissolves and Coagulates, which purifies
and fecundates; do not apply yourself to anything
else but to give your matter the requisite cooking,
without becoming impatient at the length of the
time, otherwise you will perform nothing.

Observe that the terms they employ such as; to
dissolve, to tinge, to whiten, to calcine, to cool,

case you will leave it for a longer period so that
it may become so, and the artist, by his industry,
could supplement this lack of heat, but with much
wise precaution.

The Phases through which the Substance
passes during the time of its fermentation.

The preparation is composed of four parts; the first
is the solution of the material into mercurial
water; the second is the preparation of the Mercury
of the Philosophers; the third is corruption; the
fourth is the creation of the philosophical sulphur.
The first is made by the mineral seed of the earth;
the second volatilises and spermatises the body; the
third makes the seperation of the substances and
their rectification; the fourth unites and fixes,
which is the creation of the stone.

Philosophers have compared the preparation to the
creation of the world, which was first a mass, a
chaos, an empty earth without form and dark, which
had nothing in particular but everything in general;
so that by the first digestion the body is
dissolved, the conjunction of the male and female
and the mixture of their seeds is made; this is
followed by putrefaction and the elements are

resolved into one homogeneous water. The sun and the moon are eclipsed in the head of the Dragon, and the whole world at last turns and re-enters into its ancient chaos and dark abyss. The first digestion is made as in the stomach, by a low heat more appropriate to corruption than to generation.

In the second digestion the spirit of God is carried upon the waters, light begins to appear, waters are separated from waters; the moon and the sun reappear, the elements come out of chaos and constitute a new world, a new heaven and a new earth; the young crows change their feathers and become doves; the eagle and the lion are re-united in an indissoluble bond.

This regeneration is made by the fiery spirit, which descends in the form of water to wash the matter from its original sin and to carry the golden seed into it; for the philosophers water is a fire; but direct your attention so that the separation of the waters is made by weights and measures for fear that those that are under the heavens do not drown the earth or that in lifting too great a quantity, the earth is not left too dry and too arid.

The third digestion furnishes a warm milk to the new born earth and infuses into it all those spiritual virtues of a quintessence which binds the soul and body through the medium of the spirit. The earth now hides a great treasure within its bosom, and begins to resemble the moon and afterwards the sun; take note here that in the Hermetic philosophy, the moon signifies silver, and the sun, gold; the first is named earth of the moon, and the second earth of the sun; they are born to be united in an indissoluble marriage, because neither of them fears the greatest heat of the fire.

The fourth digestion attains all the mysteries of the world; by it, the earth becomes a precious ferment, which changes all into perfect bodies, just as yeast changes all dough into its own nature; it has acquired this property in becoming a celestial quintessence; its virtue, which emanated from the universal spirit of the world, is a panacea or a universal medicine for all the maladies of creatures which can be healed. This secret fountain of the Philosophers, in which you make your matter ferment, will give you this miracle of art and nature simply by a repetition of the first work.

The whole philosophical process consists of the solution of the body and the congelation of the spirit, and all is done by the same operation. The fixed and the volatile are perfectly mixed, but this cannot be done if the fixed is not first made volatile; finally they are united and by reduction become absolutely fixed.

By these means, the superfluities of the stone are converted into a veritable essence; but he who should separate anything from our subject, knows nothing of the philosophy, because all that is superfluous, unclean, feculent, in fact, the whole substance of the composition is perfected by the action of our secret fire.

This information should open the eyes of those who, in making an exact purification of the elements and the principles, are persuaded the one should take the subtle and reject the gross; they do not know that the fire and the sulphur are hidden in the center of the earth and that it is necessary to wash it perfectly with its spirit in order to extract its balm, the fixed salt which is the blood of our stone; here you see the central mystery of this operation which will not be accomplished until you have made a suitable digestion and a slow distillation. The operative principles which are also called the keys of the work or the regimen, are four in number, the first is the solution or liquefaction; the second, the ablution; the third, the reduction; and the fourth, the fixation. By solution the bodies are reduced to the first matter and become raw again by coction; then the marriage is made between the male and female, and the crow is born. The stone is resolved into the four elements blended together; heaven and earth unite to bring Saturn into the world. Ablution is made to whiten the crow and to bring Jupiter to birth out of Saturn; this is done by changing the body into spirit. The work of reduction is to return the spirit to its body of which it was deprived by volatilization and to nourish it with a spiritual milk in the form of dew, until the infant Jupiter shall have developed the force of Hercules.

During these last two operations, the dragon, now descended from heaven, becomes furious with himself. He devours his tail and swallows it little by little until at last he is changed into stone.

Such was the dragon of which Homer speaks. He is the true image and the veritable symbol of these two operations.

Whilst we were meeting under a beautiful Pine tree, said Ulysses to the Greeks, and we were there to make the Hecatombs, near to a fountain which came out of the tree, there appeared a prodigious marvel; a horrible dragon with stains on his back, sent by

Jupiter himself, came out from the base of the Altar and ran to the Pine tree. In the branches of this tree were eight small sparrows with their mother who flew round about them. The dragon seized these with fury and also the mother who was bemoaning the loss of her little ones. After this, the same God who had sent him, made him beautiful and brilliant and changed him into stone before our astonished eyes. I leave it to the reader to interpret and to apply the moral.

SIGNS OR DEMONSTRATIVE PRINCIPLES

The colours which come upon the philosophical matter during the course of the processes of the work are black, white and red. They follow one another immediately and in that order. The beginning of the black shows that the fire of nature begins to work and that the matter is on the way to solution. When this black colour attains perfection the solution is complete, the elements are blended, the grain rots and becomes ready for generation. That which will not blacken will not become white, says Artephius, because the blackness is the beginning of whiteness and is the indication of alteration as well as of putrefaction.

The action of fire upon humidity performs everything in the work, as it does in all nature in the generation of mixed bodies.

During this putrefaction, the philosophical male, or the sulphur, is blended with the female in such a manner that they become one and the same body, which the philosophers have named hermaphrodite; this says Flamel, is the androgyne of the ancients, the head of the crow; the elements converted in this way reconcile two natures which can make our embryo in the belly of the glass and bring to birth a very powerful King who will be invincible and incorruptible. Our substance in this condition is the serpent Python, who having arisen from the corruption of the mud of the earth, must now be killed and vanquished by the arrows of Apollo through the golden sun, that is to say by our fire equal to that of the Sun.

The second principle colour is the white. Hermes says: Son of the Science, know that the vulture cries from the top of the mountain; I am the white from the black because whiteness follows blackness. Morien calls this whiteness the white fume. Alphidius informs us that this substance or white fume, is the root of the art and the argent vive of the sages. Philelethes assures us that this argent vive is the true mercury of the philosophers; this argent vive, says he, extracted from this very subtile black, is the philosophical tinging mercury with its red and white sulphur naturally mixed together in their minera; the philosophers have given it an infinity of names.

Artephius says that this whiteness comes about because the soul of the body swims upon the surface of the water, like a white cream and that the spirits are united together so strongly that it is impossible for them to depart because they have now lost their volatility. The great secret of the art is therefore to whiten the matter; so the wise artist need occupy himself solely with the dissolution of the body with its spirit, cut off the head of the crow, whiten the black and redden the white; it is this resplendent white colour which contains in its veins the blood of the pelican; let the artist abandon all those books which only embarrass the reader and engender false ideas of the work which are useless and expensive.

The process of the work should not cost more than the price of the vessel.

The whiteness is the stone perfect at the white stage; it is a precious body which, when it is

fermented will become white and full of an
exhuberant tincture which has the property of
communicating itself to all metals; the volatile
spirits having already been fixed. The new body
resuscitates, white, beautiful, immortal,
victorious; for this reason it is called
resurrection; light of day; and by all the names
which indicate whiteness, fixity and
incorruptibility.

Flamel has represented this colour in his
hieroglyphical figures, by a woman having a white
border to her dress, in order to show, says he, that
Rebis commences to become white in this same manner;
whitening first at the extremeties all round in a
white circle; the best philosophers say this sign is
the first indication of whiteness.

As the black and the white are the two extremes, and
the two extremes cannot unite except in some middle
colour, the substance when passing out of the black
does not become suddenly white; the grey colour is
found to be the intermediary because that
participates of both.

The philosophers have given this the name of Jupiter
because it follows the black which they call Saturn.
It is this fact which makes d'Espagnet say that air

follows water after it has had seven revolutions which Flamel names imbibition. The matter, adds d'Espagnet, being fixed on the bottom of the flask, Jupiter after having overcome Saturn, siezes the realm and holds the government; at his coming the philosophical child is formed and nourished in the matrix, and, at last, being born with a beautiful face, brilliant and white, thence becomes a universal remedy for all the ills of the human body.

Lastly the third principal colour is red, which is the completion and the perfection of the stone; we obtain this redness merely by continuing to cook the matter. After the first work is compleated the substance is called masculine sperm; philosophical sperm; fire of the stone; royal crown; son of the sun; minera of celestial fire.

Most philosophers commence their tracts with the stone at the red stage, so that those who read these books should not pay too much attention to them, because they are the source of many errors, until one learns how to detect the matter of which philosophers speak, the reason for their operations and the proportions of the substances which in the second work, or the practice of the Elixir, are very different from those of the first. Although the second operation is simply a repetition of the first, it is very necessary to note that what they call fire, air, earth and water in the one, are not the same as those used in the other; their Mercury is called Mercury whether it is in liquid form or whether it is dry. Those, for example, who read Alphidius imagine, when he calls the substance of the work "red minera" it is necessary first to find a red matter before beginning the work; some therefore work on cinnabar, others with minium, others on orpiment, others with the rust of iron, because they do not know that the red minera is the perfect philosophical stone.

D'Espagnet describes the method of making the philosophical sulphur; choose a red dragon, courageous, who has lost none of his natural force, and then seven or nine virgin eagles, fearless, whose eyes will not become dull in the rays of the sun; put them with the dragon into a clear, transparent prison, well closed up, and underneath place a warm bath, so that they may be incited to fight; they will not delay in coming to gripe; the combat will be long and very arduous, until the forty-fifth or fiftieth day when the eagles begin to devour the dragon; in dying the prison will become infected with the corruption of his blood and a very black poison, the violence of which overcomes the resistance of the eagles and they die also; from the putrefaction of these bodies, a crow will be born, who little by little will raise his head, he will stretch out his wings and begin to fly; the wind and the clouds will carry him hither and thither; fatigued by being thus tormented, he will look for a point to escape; be careful that he does not find any chinks; at last, washed and whitened by a constant rain of long duration and a celestial dew, you will see him metamorphosed into a swan; the birth of the crow indicates to you the death of the dragon.

If you wish to proceed further to the red, add the element of fire, which was lacking in the white, without touching or removing the flask, but by fortifying your fire by degrees; apply its action to the matter until the occult become manifest, the indication of which will be a citrine colour; then govern the fire of the fourth degree gradually by its degrees, until by the aid of Vulcan you see blossoms of red roses, which will change into amaranth, the colour of blood; but do not stop the work until you see all is reduced to very red and impalpable Cinders. This philosophical sulphur is an earth of extreme tenuity, fieriness and dryness; it contains the fire of nature in great abundance and for this reason they have called it the fire of the stone; it has the property of opening and penetrating the bodies of metals and of changing them into its own nature; they call it, in consequence, Father of the male seed.

The three colours, black, white and red must necessarily follow one another in the order I have described; but they are not the only ones that become visible; they indicate the essential changes which take place in the substance, whereas the other colours, almost infinite and resembling those in the rainbow, are but temporary and of very short duration. They are a kind of vapour which affects the air more than the earth, which follow one another and are dissipated to make way for the three principal ones of which I have spoken.

Some strange colours which may appear are signs that the regimen is faulty and of a badly conducted work; the return of the black is a certain indication, because the crow's chickens, says d'Espagnet, must never return to the nest after they have left it; premature redness is also a bad sign, and must not appear until the end as a proof of the maturity of the grain and of the time of the harvest.

OF THE ELIXIR

Second Operation

It is not sufficient to have produced the philosophical sulphur which I have now described; for the most part, people are misled, and cease the work at this stage, believing they have brought it to perfection; ignorance of the processes of nature and art are the causes of this error; in vain they will try to make projection with this sulphur or red stone. The philosophical stone cannot become perfect until the end of the second work, which is called elixir.

Out of the first sulphur there is made a second, from which, thereafter, one can multiply to infinity, one must therefore preserve very carefully this first minera of fire, for use when required.

The elixir, following d'Espagnet, is compounded of a triple matter; that is, of a metallic water or mercury philosophically sublimed, of the white ferment should you wish to make a white elixir, or red ferment for a red elixir, and lastly of the second sulphur, all according to the weights and proportions prescribed philosophically. The elixir must possess five qualities; it must be fusible,

permanent, penetrating, tinging and multiplying; it
draws its tincture and fixation from the ferment;
its fusibility from argent vive, which serves as a
medium for reuniting the tinctures of the ferment
and of the sulphur, and its multiplication in
quality comes from the spirit of the quintessence
which it possesses naturally.

The two perfect metals give a perfect tincture
because they contain within them the pure sulphur of
nature; do not expect to find their ferment
elsewhere than in these two bodies; tinge therefore
your white elixir with the moon, and the red with
the sun.

Mercury takes up the tincture at once and can
thereafter transfer it; be careful not to make a
mistake in mixing the ferments, not to take one for
the other or you will lose all. The second work is
done in the same flask or in one similar to the
first; in the same furnace and with the same degrees
of heat, but it is very much shorter.

The perfection of the elixir consists in the
marriage and the perfect union of the dry and the
humid, so that they become inseparable and the the
humid gives the dry the property of being fusible in
a slight heat; you can make this tryal by placing a

small amount on a thin plate of copper or iron and heating it, if it melts immediately without fuming, you have what you desire.

The Practice of the Elixir

Earth or red ferment, three parts; water and air congealed together, six parts; mix together and grind to make an amalgam or metallic paste of the consistence of butter, that the earth may be impalpable or insensible to the touch; add one part and a half of fire, and place all in a flask similar to the first one, having a neck twelve inches long, and seal it up perfectly; give it a fire of the first degree to digest it; you will then make the extraction of the elements by the degrees of heat appropriate to each until they be reduced into a fixed earth. The substance will become a brilliant stone, transparent and red, and will then be perfect. Take any portion you desire, place it in a crucible, put it on a gentle fire and imbibe it with its red oil drop by drop until it will melt and flow without fuming; do not fear that your mercury will evaporate because the earth will drink with pleasure and avidity that humour which is of its own nature.

You now have in your possession your perfect elixir. Thank the Great Architect of the Universe for the favour conferred upon you, and see that you use it to His Glory and do not give this secret to any except those of high principles and strict morals.

The white elixir is made the same as the red, but using only the white ferment, and the white oil.

The Tincture

The tincture, in the philosophical sense, is the elixir rendered fixed, fusible, penetrating and tinging, by the corruption and other operations which I have described. This tincture does not consist of an external colour, the colour is within the substance itself which gives tincture to the metallic form; it is like saffron in water; it penetrates into paper more easily than oil will do; it will mix very readily, like wax with wax or water with water, because the union is made between two things of the same nature. It is from this property that it has come to be an admirable panacea for all the maladies in the three Kingdoms of Nature. It searches out the radical and vital principle which it relieves by its action of the heterogenous causes which inflict it and hold it in prison, it comes to the aid of the vital principle and joins with it to throw out the enemies; they become active together and achieve a perfect victory.

This quintessence attacks the impurities in the body, as fire evaporates humidity from wood; it preserves the health and gives force to the life principle to resist any attack of illness and to separate the veritable nutriment in food from the substance which is its vehicle.

Multiplication

We understand by the philosophical multiplication, an augmention in quantity and in quality, both the one and the other beyond all that one can imagine. That of the quality is a multiplication of the tincture through corruption, volatillization and fixation reiterated as many times as the artist may please; the second augments only the quantity of the tincture without increasing its virtue. The second sulphur is multiplied with the same matter out of which it was made and by putting in a small piece of the first according to the weights and measures required.

There are three methods of making the multiplication; the first is to take a part of the perfect red elixir and mix it with nine parts of its red water; place the flask in the bath to make it all dissolve in water; after the solution, cook this water until it coagulates into a substance resembling a ruby; incerate this to the matter of the elixir and by this first operation the medicine acquires ten times more virtue than it had before, reiterate this same process a second time and it will augment to one hundred; a third time, a thousand, and so on always increasing ten fold.

The second method is to mix the desired quantity of the elixir with its water, always being careful of the proportions of one and the other, and after having placed it in a vessel closely sealed, dissolve it in the bath and follow the regimen of the second, successively distilling the elements by their proper fires until all becomes stone; then incerate, as in the other case, and the virtue of the elixir will augment one hundred—fold the first time, but this way is too long, reiterate as in the first to increase its force more and more.

The third method is the multiplication in quantity; project one ounce of the elixir multiplied in quality upon one hundred ounces of purified common mercury; this mercury placed upon a small fire will be quickly changed into elixir. If you throw one ounce of this new elixir upon one hundred ounces of other common mercury; this mercury placed upon a small fire will be quickly changed into elixir. If you throw one ounce of this new elixir upon one hundred ounces of other common mercury purified, it will become most fine gold; the multiplication of the white elixir is made in the same manner, taking the white elixir and its water instead of the red elixir; the more you reiterate the multiplication in quality, the greater effect it will have in projection, but not by the third method of which I have spoken, because the force diminishes at each projection upon the common mercury; one cannot therefore push this reiteration beyond the fourth or fifth time, because thereby the medicine would become so active and so fiery that the operation would take place instantly; the duration shortens at each reiteration; consequently its virtue is sufficiently great at the fourth or fifth time to satisfy the desires of the artist, because out of the first, one grain can convert one hundred grains of mercury into gold, at the fourth, one hundred thousand, etc. One must judge that this medicine is like the seed of wheat which multiplies each time it is sown.

It should be observed that what is called red water is the red powder; which the first operation has produced; and that the perfect elixir or red oul is the red powder produced in the second operation; this must be understood in the same way for the white.

The Weights in the work

Raymond Lully advises us that this unique thing is one only thing taken individually, but two things of the same nature which make but one. If there are two or more things to mix together, it is needful to do it according to weights, proportions and measure. I have already spoken of those in the chapter on Demonstrative Signs, under the names of the Eagle and the Dragon and I have also given the proportions of the substances required in the multiplication. One should see from these that the proportions of the substances are not the same in the first and second work.

General Rules

Before putting your hand to the work in whatever way it may be, it is very desirable to have so combined all that there will be nothing in the philosophical books which you are not able to understand so that you may be successful in the operations which you propose to undertake. For this purpose it is necessary to be sure of the substance to be used; to see whether it has all the qualities and properties attributed to it by the philosophers; because they aver they have never named it by the name by which it is ordinarily known; one should remember that this matter costs nothing except the trouble of gathering it, and that the medicine which Philalethes, after Geber, called medicine of the first order, or the first preparation is made perfect without much expense in all places, at all times, by all sorts of people; but see there is a sufficient quantity of the matter, at least, twenty or thirty pounds.

The terms used; conversion, desiccation, mortification, inspissation, preparation, alteration, all signify one and the same thing in the Hermetic Art. The sublimation, descension, distillation, calcination, putrefaction, congelation, fixation, ceration, are in themselves different things; but they do constitute one continuous operation in the process of the work in the same flask; the philosophers have given all these names to the different things or changes which they have seen take place in the vessel; when they saw the substance exhale a subtle fume, which went to the top of the glass, they named that ascension and sublimation; seeing the vapour descend to the bottom of the glass, they called it descension and distillation.

Morien said, in consequence; all our operation consists in drawing out the water from the earth and returning it until the earth putrifies and purifies; when they perceived that this water mixed with its earth, coagulated or solidified, that it became black and stinking they then said it was the putrefaction, the principle of generation; this putrefaction will last until the matter becomes white.

The matter being black is reduced to powder and then commences to become grey; this appearance of cinder has given birth to the idea of calcination, inceration, etc. and when it became completely white, they called it perfect calcination; seeing the matter take a solid consistence, that it did not flow, it then answered to their congelation, their induration; for this reason they have said that the whole magistery consists in natural dissolution and coagulation, and in cooking by one regimen until the red darkens it. One should be careful not to move the glass or remove it from the central fire, because should the matter become cold all will be lost.

To give a fire of the first degree the belly of the flask must be placed in the earth up to one quarter; for the second degree the earth must cover it, half way up the belly, & etc.

The Virtues of the Philosophical Elixir

It is, according to the sayings of all the philosophers, the source of riches and of good health, because with it one can make gold and silver in abundance and effect a cure not only for all those maladies which are curable but also, by its moderate use they can be prevented. One single grain of this medicine or red elixir, will cure paralysis, dropsy, gout and leprosy, if taken daily during some few days.

Epilepsy, colic, rheumatism, inflammation, frenzy, and all other internal complaints cannot resist this life principle. It is an assured remedy for all affections of the eyes. All aposthumes, ulcers, wounds, cancers, fistulas, noli—me—tangere and all diseases of the skin will be cured by dissolving one grain in a glass of wine or water, and bathing the affected parts; it will dissolve, little by little, stone in the bladder; is an antidote for all poisons by drinking it as above advised.

Raymond Lully assures us that it is, in general, a sovereign remedy for all the ills which afflict humanity from the feet to the head; if the illness has lasted one month it will cure it in one day; if it has lasted a year, it will cure it in twelve days while in one month it will eliminate any disease whatsoever.

Arnold de Villa Nova says that its efficacy is infinitely superior to any and every remedy of Hippocrates, of Galen, of Alexander, of Avacina and of all ordinary medicine; that it rejoices the Heart, gives strength and energy, conserves youth and makes old people young again; in general, that it cures all diseases whether hot or cold or humid or dry.

Geber, without making an enumeration of the maladies which it will cure, contents himself by saying that it will overcome all those diseases which are regarded as incurable by the medical faculty; that it rejuvenates the old and preserves health during many years beyond the normal span, simply by taking a piece the size of a mustard seed two or three times a week, fasting.

Philalethes adds to this, that it clears the skin of all belmishes and wrinkles & etc. that it will help a woman in labour, the child being dead, simply by holding the powder to the mother's nose, and quotes Hermes as his authority; he asserts that he himself has snatched many from the arms of death who had been given up by their doctors. You will find prescriptions for its application in all diseases by consulting the works of Raymond Lully and Arnold de Villa Nova.

A Vindication of the Great Work

The Grand work of the Sages holds the first rank amongst beautiful things; Nature, without the help of art, is unable to perform it, and art without Nature cannot venture to undertake it; it is a masterpiece which borders on the powers of the gods; its effects are so miraculous, that the health which it gives and preserves to people, the perfection which it gives to all things in Nature, and the great wealth it produces in a manner wholly divine, are not reckoned to be its highest marvels.

If the Great Architect of the Universe has made it the most perfect agent in all Nature one may say without fear that it has received the same power from Heaven in regard to morality; if it purifies the body, it clarifies the spirit; if it develops compound substances to the highest point of perfection, it can elevate our intelligence up to the highest knowledge; it is the Savior of the great world, because it purges all things from their original stains and by its virtue repairs the disorder of their temperament. It subsists in a perfect tenary of three pure principles, truly distinct, but which together make one and the same Nature. It is normally the universal spirit of the world corporified in a virgin earth, being the first production or the first mixture of the elements to the first point of their birth. It is worked in its first preparation, it pours forth its blood, it dies, it surrenders its spirit; it is entombed in its vessel, it ascends to heaven all quintessencified and judges the hale and the sick, destroying the central impurity of some and exalting the principles of others; so it is not without reason that it is called by the Sages, the Saviour of the Great World and the image of the Savior of our souls. One may justly say that it produces marvels in Nature introducing into bodies a very great purity and it also does miraculous things in

morality illuminating our spirits with the most
powerful lights.

I leave the readers the liberty to supplement these
results in any manner they may judge fit and
convenient.

FINIS.

TO MAKE THE LIKENESS OF A VEGETABLE

Sloane MSS 633, page 166

Rx. May Dew, and set it to putrefy in a close glass; then distill it, and out of the feces extract a Salt according to Art. Take of this Salt 1 ounce and of the distilled water 2 pints.

What herbs you would see, take a handful or two of their Seed and powder it; And pour on it the Water so much as the water may remain three fingers higher than the Seed that lies in it, and put that Salt and distilled water of Dew and the Seeds together, into a strong glass and lute it or seal it hermetically; (N.B. The glass must be high enough to take the representation of the plant). Then set the glass two feet in horse-dung for 14 days, and the matter shall become thick; And then set it in your Chamber where is clear light. And in the night when the moon clearly shines in, when it is bad weather. Secure your bottle and so keep it thus till it is dry.

And when you please to see the flowers and plant you must set the glass in a little warm sand and so in a moment; as it were comes the plant and flowers up. And when you take the heat away the Flower and plant goes.

Finis.

Merlin

B.M. 15549 (additional M.S.)

Take one and a half ounces of lamel of steel and calcine that in a pot of earth 7 days and longer if it be need till it be very red. Then take it out from the fire and keep it clean from dust and from all manner of filth till that your corrosive waters be truly made under that form.

Take tincture of Roman Vitriol (1 part) and tincture of Salt Petre 1/2 part, and 2/3 ounces of good Vermillion and grind all together upon a marble Stone. Then put all the foresaid matters in a stillatorie of glass well covered with a blind alembick well luted (luto Sapienae), then when the lute is dry, put your vessel in balneo hot, seven days-Hoc Facto- Quench the fire and when it is all cold set that said glass in a dry furnace with ashes, and do away the blind coverature, and lute a head with alembic upon the said glass, luto Sapienae. And make first a lent fire under the furnace, till that thou think that all the faint waters be drawn by the said Alembic into a receiver. and look that the joints be well luted-Hoc Facto— Reserve that water in a Vessel of glass, and lute another receiver, or the same, to thy pipe, luto

Sapienae, and set the same vessel in the furnace and increase the fire more and more till all the good water be drawn.

And also that all the said vermillion that is in the said glass be sublimed up into the head of the said glass.

Then take the said glass from the furnace and take the water that is last drawn and dissolve in that same water one ounce of Sal Ammoniack and all that is sublimed up into the head of the said glass; take it and grind it up on a Marble Stone with the said lamel of steel that is calcined before, all together very small.

Then dissolve the matters with your water that the said Sal Ammoniac is dissolved in, a little and a little, with 2 or 3 glasses under this form.

> This is the way of smoothness
>
> No other way hath Hermes
>
> He that taketh more or less
>
> All his work is like to Bes.

Finis.

OPUS PHILOSOPHORUM

British Museum Sloane 319

First dissolve Sericon in wine vinegar distilled, to each lb; of the body put I gallon of vinegar, filter it 3 times, the foeces which remain are Terra Damnata. Distil this solution in a balneo till it be congealed into a green gum, called the Green Lyon, dry it gently. Then distill this gum in a Retort of glass, let the faint water smoak away, receive the white smoke and red fume carefully, which is the blessed Liquor; in the neck of the retort will remain a Sulphur of nature.

In the bottom of the retort will remain a black foeces, of which calcine 1 lb. or more with a strong fire, till it be white as Snow, which is our Base called Mars and our fixed white earth.

The rest of the black foeces spread on a stone and with a burning coal calcine it, and it will come into a bright citrine colour in half an hour; dissolve this in vinegar as the first and distill more menstrue thereof called Dragons Blood. Reiterate this work until all the moist parts of these foeces be brought in liquor, which put to the first, called the blessed Liquor or Green Lyons

Blood.

Second Process. Set all this in Balneo to putrefy 14
days, then separate the elements and now have you
all the fire of the Stone which before lay hid in
the foeces. Distill all these putrified menstrues
M.B. in a glass body with a fair receiver, first
comes air which is an Oil, distill this over; again
7 times, until it will burn a linen cloth being
dipped therein; then it is called our ardent Water
abstractum rectified which keep close stopped;
next will distill our Flood or Water, which will be
somewhat white which receive by itself. In the
bottom will remain a thick oil like liquid pitch,
keep the water also by itself, closely
stopped, Viz. ACID (Vinegar.)

On this black liquid water put our ardent Water,
stir them both well together, and let them stand 3
hrs., then distill it M.B. put it on again and
distill, do this 3 times, then it is called
Man's Blood rectified, for which philosophers seek.
Then put on this black matter our flood or water,
mix it well and distill off the whole till there
remain most dry and black earth which is the earth
of your Stone, keep this oil and water together
close stopped for a while. Powder this black earth
and mix it with Man's Blood and let it stand 3 hrs.,

143

then distill it in ashes with a good fire, reiterate this three times, then it is called our fiery Water rectified; so hast thou 3 Elements exalted in the virtue of their quintessence i.e. Water, Air & Earth.

Third Process. The earth remaining black and dry, calcine in a furnace of reverberation into a fine white calx, mix this white earth with the fiery water, distill all with a strong fire, calcine the earth again with a strong fire, put on the fiery water again and distill it off; dry, calcine and distill thus 7 times until the substance of the calx comes over the helm. Then hast thou a Water of Life rectified and made spiritual, and the 4 elements exalted in the virtue of their quintessence. This water will dissolve all bodies, putrify them and purge them, and this is our MERCURY and lunary Aqua Fort. Distill the Water and Oil before reserved in gentle B. and the red oil which remaineth in the bottom keep diligently by itself, for it is the element of fire and our red MERCURY. Rectify the same water again and reiterate the same work until no more of our said Lunary will remain in it.

When all our Elements be thus separated, then take the first white calcined foeces called Mars or Base, or white fixed earth, imbibe it with our Ardent

Water, refined, to cover the calx partly, put on a blind head and set it in a cold place until the calx have drunk up all the water, which it will do from 8 days to 8 days, thus doing until the calx will drink up no more, but stand liquid still, then nip up the glass, and set it M.B. to putrify 140 days without moving it until it become first russet, next whitish grey, then very White like fishes eyes which then is Sulphur of Nature flowing and not evaporating in the fire and our White Stone ready to be fermented.

Then take the white Stone and divide it into 2 parts, one part reserve for the White work, the other nip up again and set it in ashes to digest till it become red and of a purple colour. So have you the Red Stone, ready to be fermented; first weigh both parts.

Take pure LUNA, and of our MERCURY ana, dissolve it in hot ashes close stopped into a green colour distill of our MERCURY from it 2 or 3 times that no part of your MERCURY remaineth with the LUNA, then nip up the oil of the LUNA and putrify it B.M. until it shows all the colours and become crystalline white which then is the WHITE FERMENT OF FERMENTS.

Put to your white Stone one 4th. part of the ferment of LUNA, lute the glass and fix them together in a

fixatory vessel under your fire, which will be done
in 2 or 3 days.

Then imbibe it with the white oil of the Stone which
is our Lunary, drop by drop, until the same be
oilish, then congeal it again and again imbibe it,
reiterate this imbibition, and congealation, until
it will flow in the fire as wax and not evaporate
on a plate of copper Nealed, then congeal it until
it be White, hard, and transparent clear as crystal.

Lapis Albus. Then it is Medicine of the 3rd. degree
and the Perfect White Stone, transmuting all metals,
chiefly VENUS and MARS into pure and Perfect LUNA.

Likewise dissolve SOL first purged with 10 parts of
Antimony in our Lunary, as before, and when it is
not dissolved your liquor will be Citrine, rectify
our MERCURY from it 2 or 3 times. Then nip up the
oil of SOL alone and putrify it in Bal., which
likewise must become black, and must stand till it
becomes White, which then remove to a stronger fire,
without opening your glass, keep it there till it
change colours, and becomes citrine, which then
is also firment of Firments for the Red.

Then put to the other part of the Stone, which is brought to a purple colour a 4th. part of this ferment of SOL, and fix them together under your fire as before, which will be well done in 2 or 3 days.

When they are become one fine powder, then incere as before is taught, with the Red Oil of our Stone, congeal, imbibe, and reiterate until it will flow in your fire like wax, and not evaporate on a plate of copper nealed, which then congeal up till it be clear, transparent hard and red like a Ruby or Jacinth which then is Medicine of the 3rd. degree, and the perfect Red Stone, transmuting all imperfect bodies, chiefly MERCURY, SATURN and LUNA into pure SOL.

This powder must be kept in a dry or warm room in several glasses, for they are so tender, and of so oily a substance, as they are apt to dissolve in any moist place.

Lapis ex tribus consistit rebus.

Viz. ex: Spiritus sive aqua viva vita
           Animia sev media fermentu
           Corporo albo vel rubeo metallico, basi
           Sive Calie.

(Note. A hand written note below the Ms. as
follows:}

      Quod est Sericon
      Est Azoth
      Almizider
      Sal Armoniac Calcinae
      Rubrai Cerasium
      Antimonium Magnetia.

            Finis.

A Translation of a few Sentences from different

Authors quoted by Dr. Becher

in his Concordantia Chymica.

P. 188. Soloman's Song. Chapter 4; vs. 12. A garden inclosed is my sister, my spouse. A Spring shut up, a fountain Sealed.

Although the prodestant Church and the Clergy, even Dr. Luther himself explain Soloman's Song to represent Christ and his Church. I say it is false. Soloman never thought of Christ when he wrote that song, or in that same song he would not have related how many women he had in his Harem, Ch. 6. vr. 8., there are 3 score queens and 4 score concubines and virgins without number. This has very little connection with Christ and his Church, Soloman wrote that song, which is an Epic Poem in Hebrew, to celebrate the Grand Work of the L. P. with this Work he has intermixed by way of metaphore the Beauties of his beloved and favourite Sultana, probably a daughter of Pharao, King of Egypt. Ch. 7. vs. 1 & 2. smelleth a little of Bawdy, very ill adapted to Christ and the Church.

In Lady Mary Montague's Letters you find a pattern of a Turkish Love Letter in the same Eastern style, as Soloman's Song. So we will leave fanaticism to the preachers and return to common sense and sound philosophy, especially when we find expressions that are used by other Alchemical philosophers and Adepts, "The Inclosed Garden {see Flamel) is the Glass." Sister and Spouse of Sol is REGULUS OF ANTIMONY & MARS, the Spring Shut Up is MERCURY SUBLIMATE, the Sealed Fountain is REGULUS OF ANTIMONY & MARS, see Ripley Revived and Count Bernard Trevisan, and other philosophers.

Ch. 4. vs. 16., "Awake O North Wind! Come Thou South!" i.e. begin with a gentle heat: See De la Brie.

What business has this with the Church of Christ?

"Let my beloved (SOL) come into his Garden & etc." 190. Soon after the conjunction in the glass, it begins to turn black, and the second true solution is at hand, and this is a certain taken that the Work will proceed rightly.

BONELEUS. In 40 days and nights the upper part will be Black and fluid like pitch, and this is a sign that the YELLOW BODY in the bottom (SOL) is truly converted into MERCURY.

THEOPHRASTUS. Let the matter stand in the anthanor until of itself without adding anything, it begins to dissolve per se, on the superficies of the matter, and a little Island appears in the middle of the Ocean, this Blackness is the Bird, which flies by night time without wings, which Bird is converted into a Raven by the Dew, which ascends and descends almost imperceptibly, yet continually.

ALANUS. Do not think that this TINCTURE is extracted all at once or in a short time, no! But only a little and by little imperceptibly day by day blacker and blacker, until in a long time perfect Blackness is obtained.

Soloman's Song Ch. 1 vs. 5, I am black but comely, vs. 6, look not upon me (do not dispise me) because I am black, because the Sun (gentle heat) has looked upon me, & etc. (denotes putrefaction).

TURBA PHILOSOPHORUM p. 192. When temperate heat does act in humid MERCURIAL bodies, it generates blackness, whereby the Germ of your Stone is generated; and when 30 days have passed, we have seen the Redness of the Carbuncle which is our Adrop, Uziphur, and our Red Lead. (Ripley has borrowed this very same expression from the Turba Philosophorum).

Soloman's Song Ch. 2. vs. 10. My beloved (SOL) spoke and said unto me, rise up my love! My fair One! And come away! vs. 12; The flowers appear on the Earth. (my Love, my fair One is R. Alba) The Winter is past, (i.e. putrefaction in blackness is passed) the Rain is over, (i.e. the Circulating Dew or Azoth is dried up) the flowers appear, (i.e. the beautiful Colours appear before perfect whiteness is obtained).

Vs. 13 & 14. Arise my Love, my fair one and come away! O my Dove, that art in the Clefts of the Rock, let me see thy Countenance, thy countenance is comely. (The King watching his Work attentively sees the approaching glorious White Tincture, shining like Oriental pearls).

P. 193. Caesar longinus of the War of the Serpent Python (MERCURY SUBLIMATE with Phoebus; SOL in Rebus).

There was a terrible lightning between the Elements, and the water (Azoth) covered the whole Earth. Rebus and MERCURY see Noah's Flood in Ripley.

But our strong Giants (MERCURY SUBLIMATE AND THE REGULUS) continued fighting without remission, night and day, whereby the humidity was all dried up.

"Those Giants fought our wonderfully small Dwarf, (the SOL in Rebis, who is the smallest or least in weight: only the 1/7, the 1/4, the 1/10, or the 1/12th. part of the whole, according to the different proportions of different philosophers, such as Aurelia occulta, Senior, Lullius, Ripley, Flamel & etc. But our Dwarf, by a miracle of God, conquered at last, caught the Giants and bound them. (SOL fixt the whole).

ALANUS p. 191. The putrefaction of the body is the beginning of our work, and is effected by gentle heat, for this reason that it may not descend, because as often as it ascends, a separation of MERCURY from the body takes place, which must not be until the MERCURY (REGULUS) and the Anima of SOL, are united, and perfectly conjoined in One Essence, in perfect blackness.

VENTURA. p. 196. It is necessary that your putrid black matter be washed and purified, and the longer the water stands on the EARTH the more the EARTH will be purified.

P. 197. When the Matter is become White, then the Spirit (MERCURY SUBLIMATE) is coagulated with the body, you must wait a long time before that Coagulation, resembling fine pearls, takes place.

ROSARIUM. When you see that Whiteness appear, supereminent above all other colours, then you may be certain that Redness is concealed in that brilliant whiteness, then you need not take out that whiteness, but you may boil your egg, until it becomes Red, and truly Red, because the more Red it is the more it is worth. And the more it is boiled, the more it is worth.

CHIRSTOPHORUS PARISIENSIS. Therefore proceed with the Philosophical FIRE (AZOTH) until whiteness is past, and until after some Citron Yellow Colours, a Red Colour follows like Scarlet, which is the highest Tincture Alba, and although you have this Colour nevertheless let the Glass stand 6 weeks or 2 months longer without opening it, and it will become more beautiful and deeper coloured, and richer, and you are the more sure of perfect Fixation.

ISAACUS HOLLANDUS p. 108. If you work with a small FIRE especially in the beginning, the matter does better retain its own humidity (MERCURIAL).

The Matter is not congealed into a hard stone, like a glass or crystal, which is melted in a violent heat, but it must be congealed into a soft POWDER, which flows like wax over a small heat.

He that governs his matter by a gentle heat, he can come to that mystery and his matter will not be vitrified, and will always remain soluble in every stage, of the Work. But by a violent heat the matter is vitrified, indurated so that it looses its solubility except you know with a great deal of labour, to restore it.

ARNOLDUS DE VILLA NOVA. O Ye unlearned Artists! Why do you raise such a violent heat? In a violent heat the Matter is destroyed and vitrified. All philosophers have said, beware of vitrification! Because it does not belong to our Stone to be Vitrified. Therefore roast him gently in all his changes, and you will get the science, and if you act otherwise, you will not enjoy the fruit of your Labours. SOLI DEO GLORIA.

Finis.

PARABLE OF THE FOUNTAIN

COUNT B. TREVISAN

B.M. Sloane MS 3641

When Heaven had so much blessed me to impart
To me ye wondrous Miracle of Art
Command was given me to converse with none
But ye clear co—partners of ye Stone.

For men possessed of Sciences Divine
Should, like ye radiant Galaxy, combine
And mix their Lights to make ye Paths of Heaven
shine
So I, obedient to ye great command.

Resolved to search and travel every Land
The Globe had ever shown, At length I came
To golden Ganges in ye Land of fame,
And Appeleia is ye Citys name

Where dwelt a man, alas that he's no more
Rendered immortal than he was before.
A man I say whom Fate had chosen forth
To Crown him King of all ye Mysteries, ye Earth

With all her wise Inhabitants can see
on this side Heaven and Eternity.

This King had made his proclamation, he
or all the Hermetical Fraternity,

Can best explain that deep philosophy
In disputation. his Reward shall be
This Book, whose leaves are pure and precious gold
And Gold's ye Cover does ye Leaves enfold.

My courage here began to fail, but I
Soon resumed it and resolved to try
The powers of Fortune, knowing well that they
Can never bear bright Victory away,

That shun ye mighty contest of ye day
So he advised, and his advice I took
Who had proposed ye Premium of ye Hook.
And I disputed till I won ye Prize,

The fatal Gold so dazling martal eyes
Almost as much as he that centers in ye skys
Then I retired endeavouring to find
Some recreation to relieve my mind

Fatigued with study, walking in ye Fields
To see ye product lovely Nature yields
I chanced upon a Fountain did abound
With limpid Water, T'was environed round

With curious Stone, and on ye top I found
T'was covered with an Oaken Trunk for fear
Beasts should defile it, or ye Fowls o'th air
Should bath themselves or wash their Feathers there.

Upon ye bank I sat contemplating
The admirable Beauty of ye Spring
And found it closed above when lo there came
A man whom I saluted by the name

Of Venerable Priest—Pray tell me why
The little Fountain, which I here espy
Is so shut up and strongly fortified
Over and under and on every side

He answered thus, T'is terrible said he
And strange ye Virtue of ye Spring you see
Of all that burst from underneath ye ground
Its parallel is never to be found.

So it belongeth to ye King alone
who knows it well, and's by ye Fountain known.
In passing by, it always draws the King
who notwithstanding never draws ye Spring.

Two hundred eighty and two days he hath
To spend in ye inclosure of ye Bath
Which makes him young again, and stronger than

the stoutest Hero of ye Race of Man.

Therefore he caused it carefully to be
with a White Stone surrounded, as you see
wherein ye Water of ye Spring does shine
Like Silver bright, or th' heaven Crystalline

And that it might be stronger to defy
The force of an invading Enemy,
Around ye top he placed an aged Oak
which had been with an artificial stroke

Cleft in ye middle, and thereby he made
Fenced from ye Sun, a most delightful shade
Then as you see it is inclosed all
First with hard Stone and a transparent wall

Then with a hollow Oak, because its nature's such
When t'is excited and inflamed too much
It is most terrible and penetrates
Even ye hardest Adamantine Gates.

And so would vanish quite away, Alas
We were undone if it should come to pass,
I asked him whether he had seen ye King
Within ye said inclosure of ye Spring,

He answered, he had seen him entering, where
He from his entrance did no more appear
After his keepers had enclosed him there,
Until ye hundredth and ye thirtieth day

When he arose in a refulgent Ray
He at ye Gate, that is his keeper—hath
A solemn charge to daily warm ye Bath
with such a heat and in proportion so

As fire is hidden in ye source below,
And day and night no intermission know.
I asked ye colour of ye King—Behold
Said he, at first you'll see him cloathed in gold

His second garment is of silk, but black
And a black doublet on his mourining back
The next he wears are White triumphant cloathes
A shirt as White as are ye Mountain Snows

His blood was red, his flesh, not so before
was as vermillion or ye crimson gore.
I further asked him whether he had seen
The King have servants when he entered in.

He answering smiled, but answered as a Friend.
No Courtiers haer upon the King attend.
He leaves his Followers as his servants all,

They must not enter ye diaphanous wall;

And none approach ye Fountain-head but he
Who does ye Heat continual supply
And office that may easily be done,
Even by a simple and most simple one.

Then I demanded of him, if ye King
Had any great affection for ye Spring
And that for him? Again he answered me
They love and are beloved mutually.

The Fountain does attract ye King, but he
Draws not ye Fountain. Yet he loves no other,
For to ye King ye Fountain's a Mother.
My question then was; If ye Monarch was

Descended of some Ancient Royal Race?
He said, he was descended of ye Spring,
Which without adding any other thing
Had made him as he was, an honourable King.

Next I enquired, what Nobles did resort
To the other Presence-chambers of ye Court?
He told me there were only six who had
Great expectations if ye King were dead.

When that should happen they would serve no more,

But have ye Kingdom as he had before.
They now are but assistants of his Throne,
In hopes of the Reversion of ye Crown.

Then I desired to be informed, how old
The Monarch was? And I by him was told
That he was older than ye Spring, and far
Maturer than his other subjects are.

How comes it then to pass, said I, that they
Kill not ye King to bear ye Crown away
Since he's so much in years? Tho' he's so old
Says he, he can endure both Heat and Cold

And Wind and Rain and Labour, None of them
can violently seize ye Diadem.
Nor could they all should they combine in one
Murder ye Monarch to possess his Throne.

Then what succession can they hope, when he
Cannot be murdered, and shall never die?
But you, my Friend, said he, must know that those
Six of his subjects from ye Fountain rose

And such existence as they have they took
Out of ye Emanations of ye Brook,
As did ye King, So they're attracted all
By it, as things by their original.

The Fountain kills ye King and them, but then
The Fountain brings ye King to life again.
He so revived, a distribution makes,
And whosoever of ye gift partakes

Tho' n'ere so little is ye portion, he
Is in possession of ye Royalty
Equal to Kings in power and riches—then
I asked my kind informant once again,

If there were any time allotted they
Should in ye doubtfull expectation stay.
He smilled again, and told me how ye King
Without his train descends into ye Spring

Altho' it loves them too, but that it must not be,
They have not yet deserved ye dignity.
When ye King enters he is stripped of those
Which he brought in, his coronation cloathes

That were as rich as eyes did ere behold
with golden leaves and wefts of purest gold
This he bestows on his first Chamberlain,
We call him Saturn, which he does retain

Entirely forty days, sometimes two more
Augment ye number of the account before.

The black silk doublet is ye proper fee
Of Jove, ye Second Chamberlain, and he

Keeps ye possession twenty days, which done
He by command resigns it to ye Moon
Lune ye third Person, has ye fairest face
of any daughter of ye heavenly Race,

And she enjoys ye garment twenty days.
Then comes ye King clad in a shirt as white
As is ye Snow, or flour of Salt, and bright
As Ariadne in a frosty night.

The King puts off this shirt which is ye share
of stern Gradivus, ye fierce God of War
Who after forty days sometimes disdains
A Resignation, and by Force remains

Two other days to sway th' Imperial Rains
Then Mars retiring, to ye Sun gives place
Who wears a yellow vizor on his face,
But is not clear as ye Celestial Lights.

Till after 40 days and 40 nights,
And then ye Sun sanguineous appears
seizing ye shirt that crimsons all ye spheres
So flaming Hercules on Oeta stood,

Fired with ye shirt dyed in ye Centaurs Blood.
I asked th' event of all these things, says he,
The fountain Gates you then shall open-and see
To all of them, and as before they sought

And had his shirt, his doublet and his cloak.
So now his red and bloody Flesh they got
To eat among themselves ye precious Heir
Of all, their Work, and Crown of their desire.

I asked again, must they so long remain
And can no sooner some reward obtain
For service done, unless they all attend
Till ye whole Circle of their Labours end?

The answer to my question was, that when
The Glorious White, ye snowy shirt was seen,
Of ye six Courtiers, four might then possess
Themselves of Powers and Riches numberless.

But they would then but half ye Kingdom gain
Wherefore they are contented to remain
A little longer in suspence to see
The full Event and End of destiny,

Which in like manner should confer on them
Their Kings bright Coronation diadem.
I asked what doctors, or what Medicine

Was sent ye King, while he remained within?

He made me answer—that they sent him none,
No man came near him but that only one,
His Keeper mindfull to perputuate
A constant, vapourous, circulating Heat.

I asked him, Is ye Keepers labour great?
More at ye first than in ye end, for then
The Fountain is inflamed. I asked again
Whether it had been seen by many men

The world, said he, has seen it, and it lies
Self—evident to every Mortals eyes;
Yet all of them that gaze thereon do know
No more than what the outward Husk does show

Then more at large I asked, what may they do?
Those Six, said he, may purge ye King again,
Three days he in ye Fountain shall remain
According to th' contents it does contain

In circling round ye place, On ye first day
He gives his doublet, next his shirt away,
And on ye third his bloody Flesh. Said I
Tell me ye depth of ye whole Mystery.

To which he made no more than this reply;

I now am tired so long with answering thee.
Which I perceiving had no more to say,
But, waiting on him as he went away

A thousand thanks I gave, a thousand more
Were ready from my unexhausted store.
He was a reverend man, so wise that even
The Astral Orbs, and wheeling spheres of Heaven

Obeyed him; all things before him shook
And trembling bowed at his Majestic look.
Now I with sudden drowsyness opprest
Beside ye Fountain did intend to rest,

And sitting on it, I c ould not forbear
But I must open all th' apartments there
In ye mean time I did so often look
On my reward, ye golden leaved Book,

Its Heaven—born splendour did so much surprise
And overpower ye vigilance of my eyes,
That, as brofe, it did my head oppress,
It so augmented now that drowsyness

That my said Book by inadvertence fell
Out of my hands into ye little Well,
Which much afflicted me, because I thought
To keep ye Prize my disputation got.

I looked into it, but alas, no more
Could I see ye Book I had enjoyed before.
Believing therefore that my Volumn fell
Into ye very bottom of ye Well

I did attempt ye watery source to drain,
So that then parts should with a tenth remain.
And when I went to draw it all I saw
It was so viscous that it scarce would draw.

While I was toiling thus industriously
I spied a Tribe, whose coming hindered me
From draining more, yet ere I left it. I
Shut all ye Fountain round, for fear that they

Like wicked thieves should steal my Book away,
But Fire was then enkindled round ye Spring
To warm ye Bath wherein to wash ye King.
I for my crime was hurried thence away

Full forty days I in a Prison lay,
When they expired I was release, and then
Returned to see my Fountain once again;
Where there appeared thick foggy clouds, as I

Have often seen hung round a Winter sky
Which lasted long. But in ye end I found

Without much labour all my wishes crowned.
But t'is no labour, you will surely say,

If choosing right, you never turn astray
In paths erroneous and ye crooked way.
Let your endeavours always be to trace
The steps of Nature in her wonted Race,

Then you ye lovely Queen shall in your arms embrace.
Therefore concluding I pronounce that he
Who in my Book ye secret cannot see
Must never hope to compass his desire

By manifold Experiments of Fire.
My Pity and Compassion move my heart
For those that wander in ye precious Art.
Therefore to them I have revealed it all,

And proved ye Operations natural.
For this my Parable ye whole work contains
In Practice, Colours, Days and Regimens,
Ways, dispositions and continuance

Till Fate and Heaven conclude ye Mystic Dance.
To end then this my Book, I pray that God
Who in ye Heavens has fixed his grand aboad
And who alone commanded me to write

Would thence impart an intellectual Light
To searching Tyros, who have hearts upright
And minds sincere, To them there shall remain
Nothing too hard, provided they abstain

From dreaming Fancys and ye subtletys
Of cheating Sophists, who by surprise
Like Mountebanks impose on vulgar eyes.
The Way is natural and but only one

Which I have in my speculation shown
I bid you all farewell in Christ, and be
Mindful of those that sink in poverty,
While Treasures unexhausted you possess

Whom ye peculiar Hand of Heaven does bless
With riches equally and happiness.
Pray then to God to send you down a Ray
Out of ye Fountain of Eternal Day.

FINIS.

THE PRACTICE OF PHILOSOPHERS

a n o n y m o u s

(WRITTEN IN BACSTROM'S HAND)

OF CHANGING WHITE INTO RED

Now let us divide the White body fixed into two parts of the which one must be conserved for the white Elixir.

The other part which is for the red put it in his glass and pour upon him of the water conserved for him, that it may overcome Him, Shut the vessel strongly and seeth him with soft fire till he liquify as water. Then decoct him further with a little augmentation of fire till he wax thick like fat or oil. After that with more fire continue till he be coagulated into a dry red stone. Then imbibe it as before is said and seeth him and coagulate him, and etc. till he be dry. So do until he have drunk all his vapour conserved for him. Put that part so dried into a round earthen vessel, glazed well without, lute it and put it in the furnace fourty one days and nights, so that he remain continually in one equal heat more than before, till the Spirit enter the body by his regimen.

Sow thy purest fruit in thy mercury till he be dead in him. Dissever the quick from the dead, the dry from the moist, with Care, and imbibe the dead with the quick, and the dry with the moist, till the dead have overcome the quick.

> Take hard, heavy, hot and dry
>
> Do together, for so did I
>
> Take hard and moist & wet
>
> Do together even I mete
>
> Then are thou richer than the King
>
> Unless that he have the same thing.

First you should put into the said body a little water and a little of your powder. Then take another glass and put a little of the same water in this glass and of the powder as you did before into the first glass. And when it is all dissolved, put it into the first glass. And so do little and little unto the time that all the powder be dissolved.

And when it is all dissolved and put in our body, then lute a blind head upon the mouth of that body Lute Sapientae. And when the lute is dry, set that same body with all the said matters in a very hot balneo for seven days-Hoc Facto.

Take up that body and do away with the blind head, soft and fair, that none of the lute fall into the body of the glass for hurting of the work. Hoc Facto. Lute a good head of glass with Alembic to the mouth of the said body-luto sapientae- and set him in a dry furnace with hot ashes; Sicut Primus and make first lent fire till that you see the said body of the said glass begin to wax red. Then increase the fire more and more unto the time that your feces that lyeth in the said glass be dry, then let it cool. HOC FACTO.

Take it from the furnace again and grind it upon a marble stone as you did before with that, that is sublimed in the head. And dissolve it in a new corrosive water made of Ti of Sal petre, a half part, and one ounce of Sal armoniac, drawn as you did before, and ever remember that in all your drawing of your corrosive waters that you put away your faint waters, that is to say, the first water, the which what he ipetive like the colour of whey. And keep it in a glass, for it will serve for another work and not for this. And all your corrosive water must be made under the form before said.

Then put your faeces so dissolved in the said body of glass and in your corrosive water in balneo again seven days as you did before. And then congeal him and dissolve him unto the time that you have him in an oil.

Then put that oil in a glass well closed and luted, luto sapienae, keep it by himself. Then take one ounce of fine Sol and make thereof thin plates.

Then take a great cruse (crucible?) and put therein as much lead as you will and melt it. And then lay a broad thin plate of Iron full of small holes upon the said cruise, and upon the same plate of Iron lay your thin plates of SOL, plate by plate, so that none of them touch each other (Stratum Super Stratum), and turn them often times.

And look that the said cruise with your lead may stand very hot and liquid. For the fervent heat and the spirit of the hot lead will cause the said plates of Sol to break into powder. Then when it is so brittle that it is somewhat brought into powder, grind that powder of Sol upon a marble stone very small, but not hard for grinding of the stone amongst the Sol. Then take that small powder of Sol and dissolve him in the corrosive water, before said, made with Sal ammoniac, sal petre and roman vitriol. And congeal him and dissolve him until the time that it be an oil. Hoc Facto.

Put the first oil and the oil of Sol together and put them in a body of glass with a blind alembic well luted-luto sapientae; and set him in balneo hot 5 days natural; and then let it cool; Then take it up and set it in a circle fire, not too hot 5 days, and then put him again in balneo another 5 days, gradus primus, and then put him into a circle fire another 5 days, sicut primus, and so do as it is before said.

What in balneo and what in circle fire until the time that the said matters be congealed into a stone, the which will be done within 9 weeks.

Then take one ounce of the said medicine and cast upon 10 ounces of fine lead molten in a crucible and one ounce of that lead so arrayed cast upon 10 ounces of fine Luna well purged, while that it doth run, and it shall be as good Sol as any in the world to abide all manner of examination and Test.

To fix Lune unto the weight and quality of Sol, take lune and fine it with Saturn as well as you can, till it be fined at the full: Then take Saturn and set him in a test one ounce or as much as you will and dry it and in the drying cast therein powder of glass, and always as the powder waxeth white strike it off, and cast in more, and so 3 or 4 times till the glass have his colours. Then strike clean away the glass from the saturn with some instrument so that the fourth part of thy saturn be washed away. Then take that saturn from the fire and in a clean pot melt him by himself and cast it into a pot with Vinegar and so do 9 times: and then is thy Saturn well prepared.

Then take 2 parts of that Saturn prepared and 1 part of the lune aforesaid well fined and set him in the fire and melt him Together, and cast him into strong Vinegar, and so do 9 times; then set it in a test and dry him till you have the lune clean by himself and then hast thou thy lune as dense and as heavy as Sol.

Now if you will make thy lune fix as is Sol to abide cementation: Take of thy Saturn prepared and melt him and quench him in strong Vinegar and so do 9 times. Then take of thy dense lune two parts and melt him and when it is melted cast in 2 parts of thy Saturn last prepared and have ready by thee a Vessel with two parts of Vinegar and one part of Sal ammoniac dissolved in the same Vinegar; And in that liquor quench thy metal and so melt and quench 9 times, and when thou hast done so, then set him in a test and fine out the Saturn till thou have thy lune clean by himself, and at all times look that thy test be right hot or thou put thy matter therein.

This lune will abide cementation Royal, like as Sol will and is called AUREUM ALBA.

Then take one ounce of Sol and one ounce of the said aureum alba and 4 ounces of Venus and melt them together and plate the metal into thin plates and lay them in cementation in a crucible with these materials:

Take Sal Alkali, sal ammoniac, Vitriol, borace croufer or crosfer is used.

------ 'tis quenched in the galls of a bowl; Red powder of tiles Sal EVEN (an abbreviation D.H.) the juice of celedone, of earth and like meche.

Then temper him with honey and make him like a paste and then make thy cementation with the plates between 2 crucibles and lute him fast, and dry him well and put him into the fire 24 hours.

Note well that thou cement him till all the Venus be wasted and have no more but the weight of the 2 principle bodies. Then melt him with borace and thou shalt have the fairest Sol that ever thou sawest, the which will abide cementation Royal. Exphat ars ista vera et pbata.

A good way for Sol: Take one ounce fine Sol and one ounce fine lune and 1/2 ounce of fine Venus, melt them together and wash him in stale wine (urine) after that you have cast him in an ingot. And beat him into thin plates and keep him to the cementation.

Then tinc. 1 part of Roman Vitriol and make powder thereof and temper him with child's urine and boil him together till thou see him rise up green and yellow, then take him off the fire and grind him with tinc. 1 part of Sal ammoniac and one ounce of Sal gemme and temper him with strong Vinegar unto the time that he be as thick as paste, and then tement the plates with that paste and melt the metal and cast him in an ingot and thou shalt have good Sol and gentle in working.

Q.D.R.

The SECRET FIRE OF THE PHILOSOPHEBS

No Author or Date

This, as it is the highest Mystery of Nature, so it is the greatest Secret of the Philosophers: The fire of the peripateticks is dry; but that of the hermetic philosophers is moist.

The common people calcine and burn with a culinary fire, we with a clean and Chrystalline Liquor; for our fire is a sulphureous Water, and the Spiritual Seed of Sulphur, which is our Mercury: So with the mercurial moistness is the Matter of our fire. As the kindled oil in the lamp is the matter of the Light, and by such a matter is augmented.

Nothing is so dark and obscure as the fire, and nothing is more occult than the manner of odering and governing it.

Pontanus after he knew the true matter (as himself confesses) erred above 200 times, before he could perfect the work, because he knew not this fire.

The knowledge of it is so necessary, that when on a time, a Society of the Sons of Art met together on purpose to debate on the Great Work so that they should manifest not everyone knew of it.

After various opinions were brought forth and conversed among them; the Youngest of the Company being asked what he knew of that Mystery, answered That he understood the fire and its Regiment, whereupon all rose up and gave him the Pre—eminency, as the Chief Master of this secret, for he can easily perform all things who knows it, and without its knowledge nothing can be performed, which seeing it is so, you may reply to me, that I have given you no satisfaction in any wise, which hitherto I have written to you.

Yet my dear friend, the things which I have said, do greatly conduce to Your desired end, for whosoever is well furnished in other things, and hitherto instructed, is well adapted for the finding out of this secret fire, which he will probably obtain, if only he continues his Inquisition, & God Vouchsafe to bless him.

I must confess that hitherto I have wrote to little purpose, unless you seem to have known the things before, however it will be a great comfort to you, that what I have written may confirm your opinions, and beget a confidence in you how more like a philosopher, than a parrot you have written, and therefore they cannot but be an incitement to promote your beginnings and progress, to which I heartily wish a most happy event; because I suppose you to be a man very fit to be made an adept. And I seriously propose, had I been so fortunate to have had the liberty granted me by my patron, which many Adepts have, I would have opened to you the whole Secret.

But my good friend, let me use that Liberty of speech as to say; that if without the tremendous Anathema of the philosophers, I might open to you the Great Arcanum viz. their secret fire, I do a little doubt whether by this means I might be an instrument of greater felicity, to you.

It is worthy to be noted that sentence of a Chief Father of the Church, God, in Mercy, denied many things which he grants in his Anger; for very many gifts of the Infinite Diety, are made rather punishments than benefits.

My most worthy friend, I heartily pray that if you go about the Great Work, and finish it; this may not be prophetical of you, the circumstances considered which accompany you, therefore I would that when soever God shall make the matter of the Hermetick

Wisdom, you would employ it about Medicine and Philosophy; but not making of Gold.

FINIS.

RAYMOND LULLY's TESTAMENT

The Final Conclusion for Understanding Raymond Lully's Testament or Codicill and his other Books, and also the Argent-Vive on which the whole intentional Intention does Depend, which is otherwise called RAYMONDS REPERTORY OR INVENTORY.

But our Secret Philosophical Water is compounded of three Natures, and it is like to a Mineral Water, in which our Stone is dissolved, and therein it is terminated, whitened and rubified. For it is not joined to the work, unless essentially moistening the parts of the dissolved Stone, whose Phlegm preserves the whole Work from combustion (or burning), by the means of the Artists Industry.

But know that all its substance, that is the phlegmaticall, is by decoction separated from the whole Compound, but our Phlegm is the middle substance, and the first Water of Mercury, in which the beginning of the Stone is, that is, its dissolution, neither does it enter with it, although they embrace one another with the Bond of Nature, unless as a Phlegm moistening the parts of the things, and not generating nor augmenting; Whence the radical moistures are the essential parts of the

Stone imbibed in the things themselves, of which alone the parts of the things consist, therefore by it, it is augmented and nourished. But it is the truly germinating (blooming, sprouting, or growing) Nature, which the more it is decocted with the Phlegm; in its Vessel, by a Fire forcing (or stirring) is the more ingrafted into (or fastened with) all and every one of the Parts, it is adorned, and so made fitter that manifold Fruits may be generated out of it: For this is called the middle nature, and the Stone, Mercury, Arsenick, and the noble spirit partaking of both extreams, the white Sulphur and the Red, binding up Mercury and converting it into better Silver than that of the Mine. But the Phlegm in which is our Sulphur is decocted, which is called Gold, is that in which the humid Air in the substance of Mercury is spiritually intercepted, until its middle part also be condensed into Water, by the means of the Action of Heat of the Body. For the Metallick Body as an active Virtue works like a male, and this because it is proper to it, because its Heat is inclined in a humid, which when it enters into the pores of Cold Mercury it is passive and is altered by little and little, and so also that which is digested by a natural Heat, it is totally (even in the smallest particles) dissolved into Water, because it is the property of every Element to have in itself a contrariety and to act on its contrary; but the parts which are not dissolved into water are in the form of an Argent Vive Amalgamated with a Body, which cannot be dissolved unless they be more subtilized by sublimation. And therefore Lute (or close) if you would dissolve it by coldness, than the more coagulate it. But the Heat of the Body does indeed digest and dissolve Mercury, as well as is condensed into water, with its parts of Mercury, otherwise dissolved. And so our Phlegm of which essential substances it consists, which Essences of the Elements are indeed mutually so directed, so that the most of each of them, with the most of the other, and the least of one with the least of the other, is altering one another, by (or with) their qualities. Whence Alexander says: That they make a mixture of the Elements, because the Elements work some certain Effects better in mixture, than simple. And this by reason that in such mixture the proper Essence of the Elements do remain, which are the proper operative Powers, as appears by our Philosophical Water, which is called metalline, because it is generated out of a Metallick Kind only.

For in this water by Sulphur then is the middle disposition, which is between softness and hardness by the (means) mediation of which, there is made a passing from the softness of Argent Vive, to the most perfect hardness of a metal, whence I say it is made by a determinative dissolution, which partakes of the two contraries: Nor is this an essential disposition of Argent Vive and Sulphur in their substance or Nature; but only a part of them; form whose alteration of the Essential Parts, they come unanimously (or together) into an earthly middle substance.

Therefore it receives a certain fixation, as a Spirit of fixation, the nearest to the Rest (or repose) of fixation, because when it is fixed, its phlegm is separated by desiccation from the whole generated Compound.

Therefore to receive (or get) the phlegm, condense Air by the hot and the dry, and rarify the dry, and you shall have the hot and moist; the Aeriall vapours is spiritually received in the matter of Gold, and of other things mixt, and therefore they are the sooner dissolved into it. But our Water after its separation by the Alembick is in Colour clear as common Water, and in the beginning of its Operation by the filtre, it is pale in Colour like the height (or tops) of the Chaff (or rather the Beard) of Wheat, the Cause of which is the digestion converting the Body to citrinity, and before it is separated by the filtre it is black; the Cause of which is the Terrestrial Corruption of the obscure Body covering the brighter parts of the Spirit in a corrupt form, although it be not so.

And after its compleat decoction, which is done by rarefaction, it is hiddenly White, which whiteness indeed is nothing else but the Spirit let loose from the body and carried by the phlegm; But the phlegm is varied in colours; but as often as the Spirit is carried so often also is the phlegm, because the phlegm is not Varried but in form. For the Spirit of every decoction is varied in Colour, and certainly the matter also is augmented in weight, in Virtue, and in power.

For when the black is imbibed with water after separation by the filtre, the black matter grows white upon the Porphory.

But the substance of this Water is the bright and illuminating Spirit of the Bodys, which in the Book of Nature was united to the metals and coagulated by fire, that is by the mineral Virtue; and when it is coagulated it is found (to be) partaking of two Natures and extreams, for it is partly fixed and partly Volatile, and therefore it is called the medium in which the fixative (or fixing) Virtue is, and while it is fixt and follows (or takes after) the proper nature of a metal. For this is called Sulphur, and a hard liquor sublimed after the manner of a Spirit into a Powder, for there can never be made a liquability in any Spirit. For this is generated from the substance of two contrary moistures, as to heat and coldness and of one kind as to their Root, the one of which is SOL, and the other Argent Vive in one degree. But in another degree it is LUNE and Argent Vive; yet this Argent Vive differs in a certain degree when it is joined with SOL, since it is not argent vive in its own nature, because it has other stronger Virtues, which are those of Heat.

Argent Vive as a passive matter is joined with SOL as an Agent; but SOL is not joined with Argent Vive, as the male with the female; but on the Contrary, because in reality the middle nature of Argent Vive has dominion over the matter of gold, and corrupts it, as if it acted upon it like a male; but it not only corrupts it, but altering it enters into the fixt and hot humidity of SOL with the Spirit, (its Spirit) and this by the mediation of that Spirit in which the Bud (the Sprout or Growth) if both of them is carried. Whence SOL is not joined to Argent Vive, nor Argent Vive to SOL, but by the middle substance, nor does the one love the other, but by such a disposition akin to nature; of which one substance, that is, that of Argent Vive exceeds the substance of SOL in a four-fold part (or proportion) because it is colder than SOL; because the active Powers are greater in SOL than in LUNE in the beginning of the Confection of (or making) the Stone.

Yet the quantity of the matter of SOL does not in truth exceed the price of one piece of Gold; but it exceeds more in the weight of the LUNE matter, because it is of a more terrestrial nature and of a weaker Power, and it is therefore strengthened with the nature of gold, and without it, it is weakened. For LUNE, this according to all (Authors) it be cold, yet it is hot in some certain degree, but it has not so great a power, that it can act on the passive matter of Argent Vive, unless it be assisted by a more active matter, and therefore the Earth of Sulphur is prepared with Arsenick, as argent vive with our Sulphur. But the preparation of Sulphur is, that if it be boiled with its Oil, with a little of the ferment, it will within 8 days return into a most red Powder tinging Silver into Gold.

Finis.

Tractatic de Summum Bonom

(extract)

B.M. MS. Sloane 631, f.183

This Epistle was written and sent by the Bretheren of the R. C. to a certain German, a copy whereof Dr. Fludd obtained of a Polander of Dantzich, his friend, which he since printed (in Latin} at the end of his tract intituled Tractatic de Summum Bonom.

Venerable and Honourable Sir.

Seeing that this will be the first year of thy Nativity, we pray that thou may'st have from the most high God a most happy entrance into, and a departure out from thy life, and because thou hast hitherto been with a good mind and constant searcher of Holy Philosophy, well done Proceed. Fear God; for thus thou may'st gain Heaven, and get to thyself the most true knowledge. For it is God who hath found out every way and it is God who alone is Circumference and Center.

But draw thou near, Listen, Take this to thee. For he who increaseth knowledge increaseth sorrow; Because that in much knowledge is much grief. We speak by experience. For all worldlings and vain glorious, vaunting boasters, gorgious men, talkers. And vain people do unworthily scandalize, yea and curse us for an unknown matter. But we wonder not that the ungrateful World does persecute the Professors of the true Arts, together with the Truth itself. Yet for thy sake we shall briefly answer to these questions: via. What we do? what can we do? or whether there are any such as we?

In John therefore we read that God is the supreme Light and in Light we walk, so that we exhibit light (although in a lantern) to the world. But thou man of the World that deniest this, thou knowest not or seest not. It behoves thee to know that in thy vile body Jesus dwelleth, this thou hast from the Apostle, and Jesus knew all their thoughts to whom, if thou adherest thou art at length made one Spirit with Him, and then being out of the same nature, who prohibiteth thee with Soloman to know as well the wicked as good contentions of men. And this thou may'st take from me out of the promises.

And hence is it that we do not answer to all; Viz. because of the deceitful minds of some. For whosoever are alienated from God, are contrary to us. And who is so foolish as to permit a newcome stranger to enter into another mans house. But if thou object that this union is only to be expected in the World to come, behold now in this thou showest thyself a Worldling, who extinguisheth light by thy ignorance. Also thou art not ashamed to make the Apostle a liar. In whom these things are more clearly manifested in these words: "So that you may be wanting in no Grace, expecting the Revelation of our Lord Jesus Christ." But thou say'st that this is not to be understood of this inferiour life, what therefore does the following verse intend. "Who shall confirm you even to thy end-for in the Kingdom of God there is no end, Therefore in this Temporal State will appear the Glory of the Lord and Jesus Glorified."

If anything is further demanded concerning our offices, our endeavour is to lead back lost Sheep to the true Sheepfold. You labour therefore in vain, O Miserable mortals, who enter upon another way than that as the Apostle wills, by putting off the Tabernacle, which way is not walked in through dying according to vulgar opinion, but as Peter willeth, where he saith as Christ hath taught me; viz. when he was transfigured in the Mount, which deposited or laying down, entrusted for safe keeping, if it had not been secret and hidden, the Apostle had not said "As Jesus taught me", neither had the Supreme Truth said, Tell this to no man. For according to the vulgar way, vulgarly to die was known to all men from the beginning of the World. Be ye changed therefore, be ye changed from dead Stones into Philosophical living Stones.

The Apostle shows the way when he saith, "Let the same mind be in you which is in Jesus." Also he explains that mind in the following words; viz. Whereas being in the form of God, he thought it no robbery to be equal to God. Behold these things, Oh all you that search into the abstruse secrets of nature. Ye hear these matters but you believe them not, Oh miserable mortals, who do so anxiously run into your won ruine, but wilt thou be more happy, oh thou most miserable, wilt thou be elevated above the circles of the world, Oh thou proud one, wilt thou command in Heaven, above this earth and thy dark body, Oh thou ambitious man, wilt ye perform all miracles oh ye unworthy? Know ye therefore ye Stone rejected of what nature it is? But thou Oh Brother, harken, I will speak with St. John, that thou may'st have fellowship with us and indeed our fellowship is with the Father and with Jesus.

We write unto you, that ye may rejoice, because God is Light and in him is no darkness at all. But if thou mayest come into us, behold this light, for it is impossible for thee to see us (unless when we will) in another light.

In this therefore follow us, whereby thou mayest be made happy with us, for our most fixed palace is the centre of all things, likewise it is much obscured, because covered with many names. Enter, enter to the glory of God, thy own salvation the gates and school of Philosophical love.

In which is taught everlasting Charity and fraternal love and that same resplendent and invisible castle, which is built upon the mountain of the Lord, out of whose root goeth forth a fountain of Living Waters (a river of love) Drink, drink, and again drink, that thou mayest see all hidden things and converse with us.

Again beware, but what? For thou knowest very well that nature receives nothing for nutriment, but that which is subtile, the thick and feculent is cast out as excrements. It is also well disputed by thyself that those who will love in the mind, rather than in the body, take in nourishment by the Spirit, not by the mouth. As for example it is lawfull to know Heaven by Heaven, not by Earth, but the virtues of this by the other, and if thou understandeth me aright no man ascendeth into Heaven which thou sleekest but he who descended from Heaven enlighteneth him first, whatsoever therefore is not from Heaven is a false Image and cannot be called a virtue.

Therefore Oh Brother, thou canst not be better confirmed than by Virtue itself, the Supreme Truth, which if thou wilt, religiously and with all thy might, endeavour to follow in all thy words and works, it will confirm thee daily. For it is a fiery Spirit, a glistening spark, a grain impassible, never dying, subliming his own body, dwelling in every created being, sustaining and governing it, Gold burning, by Christ purified, purged, pure in the fire, always more glorious and pure inbilating without diminution.

This shall, I say, confirm thee daily until, as a certain learned man saith, "Thou art made like a lion in battle and canst take away all the pull of the world, and fearest not death nor any violence whatsoever a devillish tyranny can invent viz. seeing thou art become such a one as thou desirest, a Stone and a work; And that God may bless thy labours, the which thou shalt receive
in most approved Authors to be read under a shadow, for a wise man readeth one thing and understands another.

Thou art imperfect, breathe after a due perfection; thou art foul and unclean, purge thyself with tears, sublime thyself with good manners and virtues, adorn thyself with Sacramental graces, make thy soul sublime and subtile, for the contemplation of Heavenly things, and conformable to Angelical Spirits, that it may vivify thy vile ashes and rotten body and make it white and render it altogether incorruptible and impassible by the resurrection of our Lord Jesus Christ. Do these things and thou wilt confess that no man hath wrote more plainly than I. These things the Lady Virtue hath commended should be told to thee from (or by) whom according to thy deserts thou shalt hereafter be more largely taught. These things read if thou wilt, as the Apostle willeth-Keep that which is committed to thy trust.

Farewell. F. T. F. In Light and C.

VERBUM DEMISSUM

by

COUNT B. TREVISAN

B.M. Sloane MS 3630

Translated from the French.

The first thing requisite in this Science of the Transmutation is the knowledge of the Matter, from whence is extracted the Argent vive, and the Sulphur of Philosophers; of which two the Sovereign Medicine is made and constituted.

The Matter from whence is extracted the Sovereign Medicine and Secret of Philosophers, is only most fine Gold and most fine Silver and Argent Vive, all which thou daily seest altered nevertheless, and moved by artifice in the nature of a Matter White and dry in the manner of a Stone, from whence our Argent Vive and Sulphur are elevated and extracted by strong ignition by reiterate destruction of the same, by resolution and sublimation, and in this Argent Vive is the Air and Fire which cannot be beheld by Corporeal Eyes, being subtile and

Spiritual which makes against those who think to obtain four Elements really and visibly separated in the Work, each one apart, but such know not the nature of things, and that simple Elements cannot be obtained by us, although we know them by their operations and effects which are found in these lower elements, to wit Earth and Water, as they are altered from a compact and gross nature, whereby they are moved from one nature to another. That Sol and Luna are the Matter of our Blessed Stone, the sayings of all Philosophers confirm, and in real truth saith our Father Hermes "The Sun is the Father and the MOON is the Mother", but great doubt is made of the 3rd. Composition, to wit what is the Argent Vive of which with the Sol and Luna our Composition is made. Which to know, it is to be noted that the Philosophers divided into two parts, First and Second. That Second part is by Philosophers divided into the White Stone accomplished, and into the Red Stone.

But because this notable Secret lies in knowing the First Part, Philosophers doubting to reveal this Secret have made but little mention of the First Part, and believe that if it had not been to prevent the Science of Philosophers would have remained as wholly false in its principles, they would have been entirely silent and have mentioned nothing of it;

Wherefore if they had not in any manner touched thereon, the Science would have in all points rested in Ignorance and perished, as being false in its terms.

As this is the beginning, the Key, the Foundation of our Magistery, without which nothing is to be accomplished, and that being unknown, ye Science would remain deceitful and false in practice. To the end that this great Secret might not remain unknown, which is the Stone, to which nothing strange or foreign is to be added, I have designed to make some mention of it, wholly, certainly and true, which I have seen and possessed, God of Truth being my witness, which I commit to the Secret Cabinet of the pious Soul, upon peril of the same, wherefore Philosophers have called this Secret "Verbum Demissum" which is to say a word left or concealed. It is then to be known that the Philosophical Work is to be divided into three degrees-to wit- the Vegetable, Mineral and Animal Stone. The Vegetable Stone properly and principally the Philosophers have attributed to this First, which is the Stone of the First Degree, of which Peter of Villaneauve (Brother of Arnold) saith in his Fosary, "The beginning of our Stone is Argent Vive or its Sulphurity which we must obtain from its gross corporeal substance before we can pass to the Second Degree."

The beginning of our Stone is, that MERCURY growing like a tree, be composed and sublimed by opening it for the volatile germ, or seed, which cannot nourish nor grow without the Fixed Tree that retains it, as the living nipple of an Infant.

It appears then that this Stone is Vegetable, as it were, the sweet Spirit that proceeds from the Bud of the Vine joined in the Work, first to a Body, fixed and whitening as is said in the Green Dream wherein after the Text of Alchemy is very notably described the practice of this Vegetable Stone to those who wisely discern the Truth, which for certain reasons and just cause I forbear to set down here.

The First Degree then of our Philosophical Stone, is to make Our MERCURY Vegetable, clean and pure, which is called by Philosophers white Sulphur not burning, which is the means of conjoining SULPHUR with Bodies, and MERCURY, truly and really, that he may be one nature, fixed, subtile, clean and united to the Bodies in their profundity, by the help of their heat and moisture, of which Philosophers say, Tinctures may be conjoined, but not Argent Vive Vulgar, that being cold and phlegmatic and destitute of lively operation, which consists in heat and moisture, but because it is in part volatile, therefore it is the medium to mix volatile spirits, and to adhere or adjoin to the fixed substances of Bodies, wherein is touched the cause of its necessity, which is threefold. The first as we are to join the two Seeds, to wit. Male and Female, they ought to be mingled with each other by a natural alliance and Love, and by a continual spongessity, in such manner that the moist of one be attracted by the moist of the other and by consequence the one be mixed with the other and that they be joined together, and for as much as those two Bodies, to wit Gold and Silver are made moist by heat digesting, dissolving and subtilizing, then they are the First Matter and Simple, and take upon them the name of Seed, which are near to generation, through the impression they receive by their simplicity and obedience to instrumental heat, equivalent and like to the natural heat of this MERCURY, forming and sealing thereto a kind of Elixir, for that the First Part of the Stone is called Elixir. The First Part then is the medium of conjoining extremities of the Vessel of Nature, in which Vessel the Spirits ought to be transmitted as they proceed from one nature to another, wherein is touched the Second Part of its necessity, for as the Stone ought to be impregnated with Spirits, it is necessary there should be therein a rententive faculty or Virtue embraced by them, to the end that they may be mixed with the

Body in the smallest parts. This Virtue is truly found in our Philosophical

*(NOTE: that MERCURY being putrified easily congealeth the MERCURY Vulgar which putrifaction is not without Sol and Luna conjoined).

MERCURY, as it is in part of a Spiritual nature, so is truly a pure Spirit, impregnated and purified from all fecculent and Terrestrial resident Spirits. I say true and fixed in one part for it contains the nature of the one and of the other Fire, which thing manifests and declares its ponticity or compunctuous sharpness, which appears in its operation, for by this Mercury mortified is easily congealed, the Mercury of the Vulgar, as the Text declares. Nevertheless it is not fixed by itself, but ought to be conjoined with Sol and Luna and be married in friendship and to the end, that which is not in it, may be fixed by those Bodies, to wit, that this thing which is composed of several mixtures together with their co—naturals, may directly fix the MERCURY of the Vulgar, for which cause no Bodies are mixed which are fixed, to the end that the compounded Fire, which is called MERCURY sublimed philosophically or First Matter, may be informed by proper Firments, as to obtain the force more lastingly, and persevere the contest of the Fire notwithstanding its asperity, wherefore, saith Hortulanus "That is not strange to it, to which

it ought to be conjoined: viz. fixed. And of this
MERCURY saith Raymond, "The Argent Vive by us made,
congealeth the common, and is more common to men
than the common, of less price, of greater virtue
and utility, and also of greater retention, being a
Gum more noble than Pearls, which converts and
attracts to its fixed nature all other Gums clear
and pure, and enables them to resist in the Fire,
and to rejoice therein, wherefore saith Morienus,
"Those who would obtain this composition without
this First part are like those who would mount to
the height of a pinnacle without a ladder, who when
they begin to climb find themselves cast down to the
bottom in misery and pain. This MERCURY (exuberate)
then is the first beginning and Foundation of our
glorious Magistery, for it contains in itself Fire
which ought to be replenished and nourished with a
great and strong Fire in the Second Regiment of our
Stone, now as well the Fire enclosed in our said
MERCURY by the First Regiment, as the Fire which
ought to be enclosed by the Second is by
Philosophers named the proper Instrument, which is
the Second thing principally required to be
understood in this high Magistery in such manner
that the Matter Volatile and fixt by heat and
congelation, with dissolution of the bodies as
Philosophers teach, This including and imprisonment
of the Fire of the Philosophers, for the greatness

of the Magistery have veiled under another name, to
wit. Sublimation or Exaltation of the Mercurial
Matter, as it is exalted in its noble Virtues, and
Sublimed in its degrees, "wherefore" saith Arnold
"Let MERCURY first be sublimed, that is, as MERCURY
is in nature low and base condition to wit, Earth
and Water, let it be brought to a more noble and
higher condition, to wit Airy and Fiery, which are
principles near approaching to this MERCURY in the
intention of Art and Nature. Wherefore when this
Mercurial Stone is thus Exalted and Subtilized, it
is said to be sublimed in its First Sublimation,
which it is convenient to sublime with its Vessel,
as saith Raymond in his Codicil in the beginning
Chapter: 2 de Vade Mecum de Mercurale Philosophorum.
We hope in God that our MERCURY shall be sublimed to
greater things which tinge it, and its Soul shall be
exalted into Glory, as being that which it behoves
yet to enter into its Mothers Belly, also it is said
to be born of the First Nativity, which hath regards
to all orders of the Chemical Earth, and the heart
of the workers in this Art shall not be frustrated
of joy, and shall tell thee, calling God to witness,
that as this MERCURY hath been by someone sublimed,
it hath appeared cloathed with so great Whiteness as
the Snow on the highest mountains, under a most
subtile, crystalline splendour, from whence proceeds
at the opening of the Vessel, so great, so sweet, so

excellent an odour, as the like is not to be found under the World, and I who speak this know that this (NOTE: The great odour of the Crystalline SULPHUR at the opening of the Glass, excelling all things in the World.)

most marvelous whiteness has appeared to my own eyes, and have handled this attenuated and subtilized, crystalline Matter with my own hands, and with my own sense of smelling have smelt this marvellous sweetness, and with great joy begun to shed tears upon the astonishment of marvellousness and sweetness, for which Blessed be the Eternal God, most High, and Glorious who hath hidden these marvelous great Secrets of Nature, yet hath vouchsafed to a few I know (most Reverend Father) that when you shall be acquainted with the causes of this disposition you will admire that a Matter so corrupt should contain in itself such a heavenly like nature. I am not sufficient to declare to you these Wonders, yet perhaps the time may come (if it be found expedient) that I may acquaint you with many particular matters of this nature, which here to write I have not obtained permission from the God of Nature.

Of this Celestial Nature, it is said in the Book of Prognosticks, certainly in Medicine, Celestial Gifts are found, but to proceed further, when you have sublimed this MERCURY, take fresh and new with its Blood or Ferment, that it may not grow old or stale. Present it to its Parents, Lune and Sol to the end that of these three things, to wit Sol, Luna and MERCURY our Composition may be made, and that the Second Degree of our Stone may begin.

The Second Degree, if thou wouldst then have good multiplication in the strongest qualities and mineral virtues by the operation of the Second Degree, and the help of Nature.

Take the clean Bodies and with them unite the said MERCURY in such weight or proportion, as is known to the Masters of this Magistery, and conjoin the said dry Matter {or SULPHUR) which is the SULPHUR of the Elements, and which is called Oil of Nature, and MERCURY sublimed, subtilized, dissolved and hardened by the operation of the First Degree in rejecting nevertheless all the residue and Faeces, that is made in sublimation, as of no value. But it is not to be understood in our sublimation the thing sublimed should remain in the top of the vessel, as it happens in the sublimation of Sophisters, but in our sublimation what is sublimed is a little elevated above the Faeces of the vessel and sustains itself and is joined to the sides of the vessel, and that which is foul and impure remains in the bottom, by Nature which desires to lose of its own accord, by a certain manner of evacuation, to the end it may be restored in melioration, in losing its impure parts that it may recover pure and better by which parts the 3rd. cause of its necessity, which is that as MERCURY is clean white, incombustible it illuminates our Stone and defends it from adustion, and keeps it from burning temporals and moderates the excessive ardure of Fire against Nature, reducing and bringing it back to a true Temperature and concord with the Natural Fire.

For the Philosophical MERCURY contains an excellency of the Fire in natural and sovereign virtue of which is to qualify against the heat of the Fire against Nature, and an amicable aid or assistance to the Fire of Nature, naturalizing that is converting itself, or making itself natural by sweetly, according to the natural Fire, which is a great increase known to few.

Wherefore here MERCURY is called the Earth or Nurse in this part, as it is the Matter or Sperm, without which the Stone cannot grow or multiply, whereof saith Hermes, "Ye Nurse of our Stone is the Earth (of MERCURY) of which Sol is the Father and Luna the Mother, it mounts from Earth to Heaven, again descends to the Earth, of which the strength is entire, if it be turned to Earth, which Earth and the two perfect Bodies, the right composition of the Philosophical takes birth and beginning."

*(NOTE: When the unnatural Fire in the Mercury and the Fire of Nature of the Stone are made one, then the Earth of Mercury is the Nurse of the Stone, first ascending to Heaven and descending again to the Earth fixed.)

Let these two Bodies then be sufficient for thee, being such as are sought and required, as saith Arnold de Villa Nova, that as the end of the state is perfection it perfects the MERCURY of the Vulgar and other imperfect Bodies and transmutes them. We must then necessarily acquire this virtue which is found where it is, now it is likely no ways better to be found than in perfect Bodies, for if in Bodies pure and fine there be not force and virtue to transmute imperfect Metals into perfect and true Sol, in vain should we seek it in the base, the like say of Lune and all imperfect Metals, Sol and Luna only are perfect, and all the rest imperfect.

To obtain then this Mercurial substance, wherewith all the perfect virtue to transmute into Sol and Lune the imperfect metals, you must have recourse to the two perfect Bodies and no other. Wherefore it is to be known, that the conjunction of these two

Bodies is the natural term of the last subtilliation and of Transmutaticn of the First Matter of Regeneration, because by this conjunction, by the first simple Matter is made generation of the true Elixir. Lune reduced into its First Matter is the passive matter, for truly, she is the wife of Sol and (he) is her husband, that is in very near affinity to each other, such is the ccnveniency between the Male and the Female

*(NOTE: Lune reduced to its First Matter is the passive matter of the Stone.)

in this kind of Art, of both which is engendered Sulphur White and Red, glutinous and congealing MERCURY certain better creation and nearer transmutation is made, when the proper Male is joined to its proper Female into one nature, and the Male joins most profoundly in the passive Matter by the subtility of its nature, and transmutes it more, converting its nature into another nature, to wit, into the nature of SULPHUR; of which conjunction saith Dastin "If the White Woman be married to her Red Husband they incontinently embrace one another, and are joined together. They are dissolved by themselves that these which were two might be made one Body. This copulation is the philosophical marriage and indisolvable Bond, wherefore it is said also, where these two Bodies are made one by conjunction, that they may hold one nature, to wit, our MERCURY, which is by some called the Ring or Soverign Bond, also it is called the Daughter of Platon, which conjoin Bodies assembled in Love. Compose then this our Stone of these three things, and not any other, for in other matters is not lodged that which so many do seek. This amalgama, or Philosophical composition thus ordered, it may be truly said of it, that the Stone is made but of one Thing. For this composition is a mixture of inestimable price and value, that is of such a price as cannot be sufficiently thought.

This is our Brass, that is our (aaa) or composition, made of the above said three things or Matter only, and then begins the 2nd. part of our most noble Stone, and the Stone of the 2nd. Degree which is called Mineral.

Now it is to be noted, that by the 2nd. Regimen or operation, the Stone of MRCURY which was born by the 1st. Operation, so clear and resplendent, is mortified, blackened and made vile and ugly, in fire, is so deformed with the whole compound, that it may revive with great victory, with great clarity, purity and force than it had first. For this mortification is vivifying, for in mortifying it revives itself.

And certainly these two Operations are so chained and interlaced one with another, that one cannot be without the other, as appears in the doctrine of Philosophers, for the corruption of one is the generation of the other.

All this business then is nothing else but to create SULPHUR of Nature and reduce the composition to its First Matter of the Metallick kind, for as Albertus saith in his book of minerals; "We must not so much alter or distance our Stone from the nature of Metals." Know then that this Compound is the substance out of which ought to be drawn the SULPHUR of Nature by comforting it, and nourishing it in joining to this substance the Mineral Virtue, to the end it may be made a new Nature stript from all its SULPHUREOUS terrestricity and corruption and all phlegmatic humidity, hindering digestion. It is further to be observed that according to the divers alterations or change of the one and the same Matter in digestion, divers names are imposed on it by the Philosophers according to its divers complexions, some have called it a coagulating pressure, some Azoc, Arsenic, others Album and Tincture illuminating all Bodies, some have called it, Philosophical Egg, for a Egg is composed of three parts viz. Shell, White and the Yolk, so is compounded our Philosophical Egg, or Body, Soul and

Spirit. Although in truth our Stone is but one thing according to Body, Spirit and Soul, but according to the divers reasons and intentions of Philosophy, is now called one Thing, and then another, which Plato meant when he said, "The Matter flows infinitely or always, if the Form stay not its flux", so is it Trinity in Unity, and Unity in Trinity, for there is Body, Soul and Spirit. There is also SULPHUR, MERCURY and Arsenick, for the Soul breathing, that is casting out its Vapours by Arsenick Works in conjoining MERCURY of which Philosophers say that the property of Arsenick is to breathe, or respire, the property of SULPHUR is to coagulate, or congeal MERCURY, nevertheless this SULPHUR, this Arsenick and this MERCURY are not those the vulgar think of which are not those venomous Spirits the Apothecaries sell, but the Spirits of the Apothecaries are those vulgar Spirits, theres are more of imperfection and corruption, to prejudice rather than repair imperfect Metals. Wherefore it cannot give perfection and incorruption to them, which perfection ought to be given by our Medium, vainly therefore do those Sophisters work, who endeavour to make the Elixir, from such venomous Spirits full of corruption. For certainly, in no other thing is lodged the Truth of the Sovereign subtility of Nature, but in the three matters above said, to wit, SULPHUR, Arsenick and MERCURY

Philosophical wherein the reparation and total perfection of Bodies that are to be purged, lodges, only all the Philosophers have imposed divers names on our Stone. Wherefore leave the plurality of names and regard only the compound, which is but once to be placed in one Vessel, from whence it is not to be taken till the Elementary Rotations be accomplished, that the force and active Virtue of our MERCURY should be nourished and not be suffocated and entirely lost, for the seeds of Vegetables in the Earth are not propagated by growth and multiplication if their force and generative Virtue be taken from them by any strange quality whatsoever. In like manner also, this nature will not multiply, or be multiplied if it be not prepared in manner of Water. The Matrix of the Mother after Conception remains shut up or the Fruit will be lost, so our Stone ought always to remain closed in its Vessel, nor any strange thing ought to be added, but only should be nourished and informed by the Formative Virtue of its multiplicative Virtue, not only in quantity but greatly in quality, in such manner must be influenced or put into the said Matter, the vivifying Humidity by virtue whereof it is nourished, increased and multiplied.

After then our Compound is made, the first thing to be done is to animate it, putting to it the natural

heat, or vivifying humidity, or the Soul, Air or
life. By the work of solution or sublimation with
congelation, and as you have made thy compound so
must thou have a certain manner of proceeding
according to the internal heat enclosed in the
matter, otherwise it may remain void of its designed
end, without Soul, deprived of its noble high
virtues, and would have no motion to generation as
other things by nature produced. The manner how to
put Fire into the said Matter, and to convert it
from disposition to disposition, from nature to
nature, that is from low to high degree. The manner
of this disposition is made by proper sublimation,
Impregnation, mortification for Resurrection and
Sublimation in the light Elements, so that all the
circle of this noble Magistery is nothing but
sublimation perfect, which nevertheless hath many
particular operations annexed, chained, interlaced
or joined together, but two are principally
attending the whole Circle, which are perfect
dissolution and perfect coagulation. So that the
whole Magistery is perfectly to dissolve, and
perfectly to congeal, dissolve the Body and congeal
the Spirit, and this operation has tyes and
alliances together, that the Body never dissolves
but the Spirit congeals, nor also the Spirit
congeals not, except the Body be dissolved. As
Raymond and all the Philosophers say, that the whole

Magistery is nothing but Dissolution and Congelation by the Ignition, of which Operation many great and learned in this Science have been deceived, thinking to understand by confidence in their learning, the Circles of Nature, and the manner of Circulation.

It is then expedient to understand the manner of this Circulation, which verily is nothing else but to imbibe, refresh or moisten the Compound in due weight or proportion with our Mercurial Water, which Philosophers command to be called Permanent Water, in which Imbibitions the Compound is digested and congealed to its natural accomplishment.

T'is most certain, that if Earth is to be made Fire, it must be subtillized or prepared, that it may be brought to greater simplicity, so is the Compound attenuated and subtilized, until the Fire bear rule therein.

And this Sublimation of Earth is made by Subtile Water, most highly sharp and penetrating, not having any feculency or Odour, as Geber saith in his Summary, "Such is the Water of our Argent Vive, sublimed and brought back to the nature of Fire, under the name of Vinegar, Alum, Salt and many other sharp Liquors and other like things, (even until now hid and covered) by which Waters bodies are subtillized, reduced and brought back to their First Matter, neighbouring Stone, or Elixir of Philosophers, where it is to be known that as the Infant in the Mother's Belly ought to be nourished with natural nourishment, which is Menstrual blood, to the end it may be increased and grow in quantity and stronger in quality, so ought our Stone to be nourished with its Fatness (as saith Aristotle) of its own proper nature and substance, but what this fatness is that is the nourishment of life, increase, and multiplication of our Stone the Philosophers have wholly concealed it, as being the greatest Secret they have sworn never to reveal or make manifest to anyone, otherwise than their writings declare, but have remitted it to the hand of God alone, to reveal or conceal as it shall please Him. Nevertheless this fat and viscous humidity, vivifying or giving Life, the Philosophers have called Mercurial Water, or oil Permanent, a Water abiding the Fire and also a Divine Water, and is the Key and Foundation of the whole Work, and this Mercurial Water

*(NOTE: the Key of the whole Work is to make the Perfect Body and Elixir, Unctuous.)

impregnated and permanent, it is said in Turba "That Bodies must be wrought by flame of Fire that they may be broken, torn and debilitated," to wit, by this Water full of Fire wherein the Perfect Bodies are so much washed as it is dissolved and made Water, which is not Water of the Clouds, or Fountain Water, as think the ignorant and foolish Sophisters: But this our Permanent Water which yet cannot be made Permanent without the Body with which it is joined, that it resists the Fire without flying, wherein our Permanent Water is the whole Secret of our Stone, for by that Water is our Stone Perfected, for that in it lodges the vivifying humidity of our Stone, as being the Life and Resurrection of it, of which our Secret Water it is said in Turba, "The Water alone does all, for it dissolves all, it congeals all that is congealable and divides and rends all without any other aid. It is that which Tinges and is Tinged," In brief, our Work is no other thing than Vapour and Water, which is called whitening and rubifying and casting off the "Blackness" of the Bodies, which Philosophers have called Permanent Water, fixed and inoorruptible and incombustible Oil, that cannot burn. T'is the Matter which Philosophers have divided into two parts, one of which dissolves the Bodies by Calcination, that is, by reducing it into calx and by congealing itself and the other part of the said Water cleanses the Body from Blackness, Whitens and reduces it, makes it fluid and running in multiplying its parts.

This Water is called in Turba, "Most sharp Vinegar, and penetrating kindly by a vivifying heat, containing an invariable Tincture, which cannot be defaced or blotted out."

This Water is named by Artephius, temperate or the Moisture of the Wise, Wine of Choleric Youth. This Water is greatly concealed by the Philosophers under divers and various names, and is known but to few people. Hermes possessed it, and handled it, Alphidius hath treated of it. Morenius hath wrote of it, Lully understood it, Arnold perceived it, Raymond hath feasibly declared it, Geber knew it, the Text was not ignorant of it. Rasis, Avicen, Galen, Haley, and above all Albertus hath wisely hidden it. Dastin Bernard de Greves, Pythagoras, the Ancient Merlin and Aristotle understood it well. In short this Water is crowned Victorious Secret, Celestial and Glorious Water, the last and final Secret for the nourishment of our Glorious Stone, without which it is never amended, increased, nor multiplied, for which Philosophers have concealed the manner of making it, as the Key of their Magistery, and in truth, I have read above one hundred books of this Art, and in none of them I found the perfecting of this MERCURY and Permanent Water and have seen great and learned men in this Science, among whom I have not found one that had this secret, except one able Physican, who told me that he had been thirty-six years in earnest prosecution before he obtained it.

Of this nature it is said that a double nature is given to it, to wit, of SOL and LUNA, in the bowels of whom this Argent Vive is multiplied, as in the proper Belly of its Mother, lodged and purged, and converted into White SULPHUR not burning in the action of the heat of the Fire, being therein regularly informed by Art and qualities of Sulphur, having been before introduced or placed in the said Argent Vive. So that this Mercurial Water is nothing else but the Spirit of Bodies converted into the nature of Quintessence giving virtue to our Stone and governing it, and the Stone of our Composition is the containing Matrix or expedient place, to wit, Mother Earth, or vessel of Nature, retaining the formative Virtue of the Stone, wherein natural heat is placed, which is formative Virtue issuing from the Vessel by the Fifth Spirit, Wherefore it is called Mother and Nurse as giving natural Virtue to Sulphur, feeding and nourishing it.

This is our Composition, wherein this natural Vessel, whereto the Spirits are transmuted from nature to nature as they proceed, and so much the more as they are transmuted and altered in this

Vessel retained so much, the remoter are they moved from their corruption and imperfection whatsoever, and the more they approach to the terms of purity and perfection, so long till they obtain the accomplishment of a Quintessence.

Wherefore they take, or are cloathed with a New Nature, which is clean, White, pure, cleansed from all corrosiveness, and phlegmatic volatility.

In such affinity then of the Vessel the humidity of the Spirit wherein the abovesaid is inclosed by its viscosity or glutinous nature is retained in adherence, or firm and natural conjunction and heat, as in its Radical Humidity mixt and mortified, and after Death revived by sublimation by a joyfull deliverance in elevating itself wholly from a Saline and Bitter Nature.

Then it is mighty to sustain itself nourish and multiply itself, as being already a kindled Fire and simple nature, which t'is convenient to nourish with Whey, to wit, its enlivening Humidity of which in part it was engendered, which is Permanent Water, Virgins Milk or Aqua Vitae coming from the Vine, for they are wholly different. It is nevertheless called Aqua Vitae for that it vivifies our Stone ot Resurrection, it is also called Blood incrudated,

White Menstruum, nourishment of the Child, food of the Heart, Water of the Sea, Food of the Dead, and Argent Vive of Philosophers depurated from feculent terrestriety by Philosophical Sublimation. After then our Compound is made, it should be placed in its Secret Vessel and decocted, or baked in a very soft heat moist or dry, and moistened with our Permanent Water by little and little, in dissolving and congealing so often till the Earth arise Filiated, which must be afterwards calcined and infinitely incerated in fixing with the said water, which is called fixed incombustible Oil, till it flows and melts soon like into wax, of which saith Raymond, "The manner of ceration is, the Stone be sublimed with part of the reserved humidity which never leaves its Body, being mixed, with circulation gives true fusion," and afterwards he said "It is commanded that thou moisten or refresh our Stone with its permanent Humidity, whereby its parts are clarified as appear, for after the perfect clearing or purgation of our Stone from all corruption, particularly the two Sulphureous humours, to wit the unctuous combustible fatness and phlegmatic vaporiety, the Stone is brought to its proper nature and substance, not burning, and without this Humidity, our Stone is never amended, nourished, augmented or multiplied.

T'is to be noted also that our Stone in digestion is moved to all the colours in the World, but three are principal, of which good care and notice are to be taken, to wit, Black colour, which is first and it is the key of the Beginning of the Work; of the Second kind or degree, the White colour is the Second, and the Red is the third, whereof it is said that the thing of which the head is Red, the feet White, and the eyes Black is our Magistery.

Note then that when our Compound begins to be moistened with our Permanent Water, then is all the Composition turned like melted Pitch and is Black like a coal, and at this time is our Compound called Black Pitch, Burnt Soot, Melted Lead, Foul Laton, Magnesia, and the Black Bird. For there is seen a Black Cloud flying in the middle region of the Glass in desirable manner being elevated about the vessel, and at the bottom, the Matter melted like Pitch, of which saith James of the bough of St. Saturnia, a blessed Cloud flying about the vessel wherein the Sun is eclipsed, and Raymond says, "When this Mass is thus Black then it is dead and deprived of Form, then it is said to be a dead Body, and out of good temperature, its Soul being separated from it, then is the humidity manifest in the colour of Black Quicksilver, and stinking, which before was dry and White, of good odour heating, cleansed from SULPHUR by its first operation and is now depurating by this Second Operation.

And for this is thy Body deprived of Soul which it hath lost with its Splendour and marvellous lucidity which at first it had, and all now is Black and Ugly, wherefore Geber names it from its properties, stinking Black Spirit, occultly White and manifestly Black, naming it Living Water and Dry.

This Mass thus Blackened is the Key and sign of perfect invention of this manner of Work of the Second Regimen of our most precious Stone, wherefore saith Hermes: "This Blessing being seen, believe that you are in a good Path, and have kept in the Right Way."

So that this Blackness in colour shows the true and right manner of working, for hereby the matter is made deformed and corrupt with a true Natural corruption from whence follows generation and real disposition in this Matter, to wit, the requirements of a new Form which for lucid serenity or clearness, beauty, purity, marvellous splendour and fragrant odour or great sweetness.

Now when the work of Blackness is accomplished we must come to the Work of Whiteness, which is one of the Roses in the Philosophers Garden, by many desired, required and expected, but as above said before perfect whiteness appeareth, all the Colours that may be thought of are seen and perceived in this work, of which care need not be had, but only to Whiteness that must be expected with great constancy. The way nevertheless of working to the Black, to the White, and to the Red is always one, to wit, bake and decoct the Compound in feeding with our Permanent Water, to wit, decoct. Compound with Red, by which imbibitions and digestions is extracted from the Stone this middle Substance of MERCURY, which is the whole perfection of our noble Mastery, in such manner that our Stone should be purged not only from sulphureousness but also from earthiness by sublimation of Water, Calcination of Earth, moistening and decocting of them by reduction, between distillation and calcination, and after conjoining with proper SULPHUR, by its measured natural heat decoct, or bake so long until it be congealed and deprived of all Sulphureous humidity by the union of natural heat and Fire thereto corresponding, and after it is sublimed into SULPHUR most White like Snow.

By this it appears that our Stone contains two substances of One Nature, one volatile, the other Fixt, which, and either of which the Philosophers call Argent Vive, because in the operation of this Stone, the Stone ought to be separated from all combustible and corrupt Sulphuriety, and that there remains only the pure subtility and middle Substance of Argent Vive congealed and

*(NOTE: Both the fixed and the volatile parts are called Argent Vive)

depurated of all external SULPHUR or strange corruption, and this depuration is made when the Body is turned into Spirit, and the Spirit into Body by reiterated calcination, reduction, and sublimation, whereby is made the dissolution of the Body, and is but One Operation whereby all these things are performed, to wit, solution of Argent Vive fixed with congelation of certain parts of the Volatile, and Ablution thereof with Water proportioned (with Fire) and the congelation of the said Water into a Stone, by the medium and operation of Heat of the Male and the Female, then truly is born the Stone after the first conjunction of them and not before, as in Man and Woman by this operation is the Body divided, subtilized and diligently governed, till its subtile Soul is extracted from its solidity and turned into a thin attenuated and impalpable Spirit. Then the Body is made no Body, and the no Body is made a Body, and this is true, and the true invention of the rule of working.

It is to be known though all Bodies are to be dissolved by a penetrating Spirit, with which it is to be mixed and whereby without doubt is made a like Spiritual, and as this Spirit is sublimed it is named Water, which washeth itself and cleanseth as above said, in rising with a most subtile Substance, opening the corrupted parts and of it. And this Ascension Philosophers have called Distillation, Ablution and Sublimation perfect and accomplished. The Stone is then vivifyed by the vivifying Spirit or natural Soul of which it was deprived in Blackness, and is now inspired, animated, resuscitated, reduced, and carried to its last end of subtility and purity, and is a Stone, Crystalline, White as Snow, rising from the bottom of the vessel sticking to the sides of it, the remainder of it resting in the bottom of the vessel below this Christalline Stone separated from its residence gathered apart and sublimed without the said residence for if you try to sublime it with its residence you shall never make separation of them, and so your labour will be lost.

Sublime it then without its residence, and it is the White Foliated Earth of the White Sulphur not burning, congealing, and after perfectly fixing

MERCURY, cleansing all foul Bodies and perfecting
the imperfect, reducing them to true LUNA.

This Sulphur so sublimed, no whiteness in the world
exceeds it, for it is divested of all corruptible
things, and is a new nature, a Quintessence arising
from the pure parts of the four Elements. T'is the
SULPHUR of Nature, Arsenic, not burning, the
Incomparable Treasure, the joy of Philosophers, and
the Delight so much desired by them, the White,
Clear and Foliate Earth, the Bird of Hermes, the
Daughter of Hippocrates, sublimed Alum, and Sal
Armoniac, the Daughter of the great Secret, and the
new White Black Bird whose Feathers exceed
Crystalline Brightness, White as Snow, of great
Splendour and most great satisfying Odour of
Sovereign Purity, of clean subtility and agilty.

This Philosophical White Black Bird is of venerable
virtue, for it is the Substance of the purest
Substance in the world which is the simple Soul of
the Stone, clean, noble, separated from all
corporeal thickness and by great subtility divested
of all bodily grossness. This White Incombustible
SULPHUR it is convenient to calcine the space of its
dry decoction, so long till it becomes most subtile
powder, impalpable, deprived of all sulphureous
humidity. Then let it be incerated with White Oil of

Philosophers by little and little till it suddenly flow like wax, and without ceration being finished, which is nothing but reduction, to fusion or melting the thing would not melt. Then is our Glorious White Stone of Philosophers finished, fusible and melting, White as Snow, of new (Verdure) greenness, perservering in the Fire, retaining and congealing MERCURY after fixing it, tinging and transmiting all imperfect bodies of Metals into LUNA, of which cast one part upon one thousand parts of MERCURY or other imperfect Metal, it shall change it into better Silver, finer, purer and whiter than that of the Mines. The manner of Projection and Multiplication of the White and Red Stone are both one, but the multiplication may be done in two manners, one by projecting one part upon one hundred parts more into pure LUNA or pure GOLD. There are other ways more profitable and secret to multiply the Medicine in Projection, wherein I am at present silent, but by multiplication the Stone is augmented without end, to wit, by Digestion, animation and Imbibition with Mercurial Oil, which Oil is also named of the nature of Metals, and this multiplication is only done by Imbibing and refreshing the Stone with the said Mercurial Oil Permanent, by dissolving and congealing so often as one will, for the more the Stone is digested the greater is its Perfection, and the more it will transmute, for it will be more subtile. And herein is accomplished the White Celestial Rose, of good odour, embraced by all the Philosophers.

When the White Stone is accomplished, you must dissolve one part of it, and so calcine it (as some will have it) by long decoction till it become like impalpable Ashes, so soft not to be touched, coloured Citrine. Then imbibe it with the Red (Red Oil) till it become Red as Coral, as Raymond saith in his Codicil in his Chapter on Calcination of the Earth. "Forget not to Calcine it in its kindled Fire."

The Matter of the Earth foreknown of thy Stone, with reiterated dissolution, distillation of Water, and calcination of the Body till the Earth remain White, void of all humidity and after a longer and stronger continuation of the Fire and imbibition of the Water till it become like a Hyacinth in powder, impalpable to the touch, the sign of which is manifestly shown.

As to its last Calcination it remains deprived of all humidity, spoken of in the second principle process in the Second Regimen which is to take the Stone Red, of which saith Geber it is not without the addition of a thing tinging (or Ferment) it, which Nature well knows, to wit, without it be

imbibed and tinged with this Celestial Water or Oil, of which says the Lily of the Philosophers, "Oh Celestial Nature, how dost thou turn our Bodies into spirits, Oh Marvellous puissent who is above all, and surmounts all, and is the Vinegar which turns the SOL into true Spirit and Luna also, without which neither Black nor White nor Red can ever be produced in our work. For without this Nature
*(NOTE: Of the Vinegar which converts SOL and LUNA into Spiritual Ferment without which the Stone is not amended.)
is joined to our Body, which converts it into Spirit as Spiritual Fire, tinging it with venerable Tincture that can never be blotted out."

This Water Hermes hath called the Water of Waters, Alphidius Water of the Indian Babylonian and Egyptian Philosophers. This Water whereby Bodies are turned into Spirits and their First Nature and Matter of our Stone, are never amended without it, but add to the White the White Water, let then the Red Stone be moistened with the Red Water to that end by long decoction or bathing, as by long imbibition, and by continual moistening, it be made as Red as Blood, the Hycinth Scarlet or Ruby, shining as a Light in a dark place as a kindled Light.

And lastly, that our Stone be adorned with a Red Diadem, of which saith Diomedes, "Honour our King coming from the Fire with his Wife and take heed of burning them by too great heat, bake and decoct them sweetly that to the end they may be made Black, then and afterwards Citrine and Yellow, then Red, and last of all Tinging Venom. For these are to be made by the division of the said Water, as said Egistue, "I command that you put not all the Water together but by little and little, and bake gently till the work be accomplished."

And so it will appear that the Stone will remain Red of a true Illuminated Redness, clear lively, melting like wax, by the Tincture whereof Vulgar or Argent Vive and all other imperfect Metals may be presently turned into true SOL much better than the Mines produce, wherein is accomplished our precious Stone, which is infinite Treasure to the Glory of God who lives and Reigns perpetually.

Finis.

ARCANA DIVINA

(The Divine Secret)

Anonymous

Published by Dr. G. A. Fuchs in Collected Volumes 1885-1916 of the Provincial Library (Vol. 8, History of Literature, p. 417), and in the Annual Report of the Communal College of Komotau (Bohemia) from a Manuscript from the Ossegg Foundation.

Forward

The Arcana Divina, as published in manuscript form, in what is to follow, originally came from the Ossegg Cistercian Foundation of Bohemia. The manuscripts consists of sheets fastened together in folio format. Since the information given by the authors, as well as their of publication, has not been strictly verified, it cannot therefore be accurately determined whether an original manuscript or merely a copy is under consideration here. The entire nature of the manuscript report, which cannot yet be ascribed to the organized written effort of individual people, or even to an actual stage of development, while other words, appended subsequently, probably appear in the manuscript due to additions in the 17th century. In order for the alchemical literature to make a contribution, the manuscript is translated freely enough that an inquiry from the Biographical Institute of Berlin involving a subsequent publication of the Arcana Divina would not be required. Critical remarks about subject matter and text of this publication will be the subject of a later investigation.

Arcana Divina

From the providence of God, on an actual basis: The source and fabric of wisdom, the Rock of Ages, from whom no secrets are hidden and at the same time, from all three Persons a universal work will be accomplished, which we will reveal, openly to you through God's love in our subsequent writings; which we carried out even as we were commanded to do.

The first man, Adam, was a universal man made in the image of God, from whom his son Seth afterwards derived children and descendants, which had knowledge of all this. So two columns were provided --- one of water, the other of fire --- both with rejoicing.

Since afterwards the flood of sins came from the Assyrians and the Chaldeans, and after Abraham's time, from the Egyptians, and finally fro Greece, such columns were destroyed in the abomination of desolation. See! In those times there was distress and there were lamentation; indeed so, but nevertheless everything will be revealed so that no longer will anything remain hidden, as though exposed by the brilliant light of the sun, with its golden rays, but those who are still able to rejoice at that very moment because of the unmerited mercy of God, will be made whole.

Oh, Almighty God, Oh Wonderful Creator, Oh Light of Unfathomable Eternity, enlighten us and send us your Holy Spirit, by the help of whom we will be able to fulfill our calling completely in Thy Holy Will and to Thine Honor and Glory, given to us in that great Mystery for the races of men who are to follow us and leave behind us, as much as it is possible, a useful record and thereby show our love for the Eternal God and to honor His Holy Name and to praise Him.

Will you then, oh man, consider that you, who should consider this, our writing, and come to realize that it is of very great value, since it will cause you to marvel at the Majesty of God and the miracles he performs daily. Moreover, where the Majesty of God is not thwarted, so that (even as in Joshua's time) the sun can be made to stand still in its daily course for a period of 24 hours and entire mountains can be consumed in flames of fire, while such shadows still surround us that we must trust God for everything. One day, however, we will soon come to understand our true worth and, submitting to God's sovereignty, we shall make a proper beginning with Jehovah and give to Him the highest Glory He merits. Thus we will write to you herewith in the certain assurance on the basis of all three realms of Nature, each in itself alone being made manifest from Universal Knowledge. In you ultimately will come to reside these highest and greatest treasures which are possible, and in which lie the basis for all wisdom and knowledge. The Creator and Maintainer of all who are in need, especially making whole all who are sick in spirit, the Liberator and Saviour of all those who are in prison, and the Overcome of all obstacles in the whole world, and the Possessor of that wisdom that comes from the Most High God, which He has had throughout all eternity and which remains in Him unchangeable through all eternity and always will, until revealed by Him.

Many, indeed almost all, of the great philosophers strive with great zeal to find a universal solution for all problems and devote considerable amounts of time, effort and expense, but as yet, in our times, they have not been able to provide such assurance as is suitable for this effort and this has resulted only in deceptive boasts from those who have attempted such things and has limited the materialistically inclined and cooled the ardor of those who saw the impossibility of demonstrating the results hoped for by the use of mercury in demonstrations using various kinds of wood, coal and lamp oil. So drastic were these results that such materials must always be treated with caution. Finally, by penetrating this chaotic mass of information, from which it was originally taken, the results indicated here were those arrived at.

For this reason we report to you and tell you that it should be understood from the beginning, under circumstances where you will be considered an expert in this matter, that even a learned philosopher or practitioner on the subject of fire, will find that working with combustible material is quite dangerous and, even more, during the preparation of such materials in the natural course of events of such things, the danger is added to and even compounded. It could be demonstrated to you just where we might want to discontinue such precautions in an effort to save time, such error will now be pointed out to you as inadvisable.

Just as the familiar mercury involves the correct method of using our apparatus and the passion for Natural Philosophy (i.e., Science) is the primary object for our art and the body of knowledge about life is of necessity to be sought after, so there must be a universal solvent which can be prepared according to our standards of measurement.

Let it now here be correctly understood (and trust that Nature will be on our side and help us) that taking the familiar mercury and putting it in a sealed receptacle in front of our apparatus, as noted in out first tabulation (index) and in several ways in our first experiments indicated at the end of our work, such as the separation of the mercury through our "magical Natural Fire" in the form of a cloud which rises to the top of the receptacle, leading to the receiver, without the passing over of the heavy material (introduced) as a continuous flow of liquid, while the active spirit (reactive gases) evolving therefrom still remains confined to an area in which further measurements can be made. This can be made now not only on metals and minerals but also upon vegetables and animal matter in aqueous solution, as well as the original material in water in which the living spirit (active principle) is retained without being distorted at all along with its entire power to grow; and also that henceforth no more can be brought, in the future, as had been obtained previously, and then it will be a matter solely of operating with our same magic and our passion for Science and can readily from time to time, within the span of three hours, produce an entirely new material, and far better than would seem probable by its creation therefrom. We will now report on our study of this process and instruct you further about its character. Here is how you should make a quantity of such a universal solvent.

Here we will demonstrate the true philosophical purpose (using the Philosophers' Stone, by the help of a permissive God and our magical knowledge of Nature). Fire must be used, but nothing else is required that is necessary for life. Astral spirits, from which all of the elements arise, are created by God and through Him is everything created that has been created, even to the end of the world, and it will be under His control, for without Him there is nothing at all, except death. Therefore, it falls to us, through the Holy Orders of God to us who dwell here below, in all the Kingdoms of the World, which produce, maintain, and even cause the increase in everything that life has here. Where, however, planting is the earthly realm, you see, it represents the plan of creation, which is no other than a purposeful idea. Thus Nitrum (crude sodium carbonate) when mixed intimately with common salt, even though impure, can be a beautiful thing, but the impurities can bring, in this instance, a heavenly appearance through our magical natural philosophy ("natural fire"). The same now gives up its effectiveness and becomes a splendid magnet with which we continue our evaluation, which subsequently will follow with more than one example, which will suffice for the present. We shall now bring out the pertinent points about mercurial properties.

How splendid and strong this love of nature is portrayed and tests the general idea of a world spirit, which is the true philosophic subject, and the only one of all of them, through which the most discerning thoughts are fulfilled, is to be seen as true by them, since a pound of mercury in pure form is placed in an open glass dish specifically selected for this purpose and for our magical passion for science (i.e., the "natural fire") as well: That the focus might not be too strongly affected and even, little by little, might tolerate a little Nitrum thereupon, so that it will then be see how sodium carbonate behaves like the universal spirit of a magnet of the world, toward the mercury which is present in the liquid state and serves as a universal standard. Some, and to a certain extent the most part, may be selected. Philosophical concepts will result only when these processes will be able to make gold and other noble metals, but not including minerals.

Three is a prima material which is considered to be the first (original) material from which the Philosophers' Stone or the Universal Tincture can be produced. This is very likely an impossibility since, as is widely but mistakenly believed, by reducing it to such a state, no other use can be obtained from it. Still, as has been previously indicated, it may even remain for a long time in a higher state, by exerting greater care in an ordinary fire. Yet it is still possible that nothing of universal value may result, but only an empty idea, the particulars of which are not worth mentioning here, that the transmutation of metals has been known to take place. When, however, such an occurrence does take place, it is of a magical nature known only to us and involves heating of the salt on our apparatus, which salt must be nitrified to a strong degree by the process of ethereal calcinations followed by treatment with nitric acid, whereby it is quite possible to obtain a thousand times greater yield than from previous methods. This volatile mercury vapor, then, can be dissolved in vitriol (sulfuric acid) with vigorous action, resulting in the transmutation of all imperfect metallic materials into large amounts of gold and silver, each of the finest quality. But know you from the outset that this vitriol (from which you will recover the gold, will be the same ferment as that from which you will recover the silver) requires that you have to have a particularly rich (i.e., concentrated) tincture, that is, there is no universal material, of long standing which will apparently be useful to you, even to a very limited extent, for without the ferment there will be no progress toward transmutation at all. So we will now examine this, very briefly, and report to you further about this.

How and in what way control of all types of medicinal materials and mineral (both pure and impure), regardless of the dosage, will be brought about will become a completely fixed standard.

We will now describe to you substances that can be both wild and domestic medicinal agents and minerals, noting thereon the weight and registering them under such crude designations as given them by the mountainous areas where they were obtained and then further classifying them to the extent that it is possible, so that it happens that when the crude earth is dried and then mixed with a little natural soda (i.e., sodium carbonate) on a clean, flat piece of tile and subsequently exposed to the magical Philosophers' Stone and then afterwards calcined and stirred with a small stirrer, while heat was applied evenly, so that when the product was finished and the calcined materials weighed, a miracle was observed. That is to say, there was a significant increase --- nay, even a large increase --- in the gross weight found, as though this magical preparation were able to capture and concentrate the Spirit of the World and be blessed by the resulting increase in its value as well. Both wild and domestic mineral products themselves, without the addition of anything except the attaching of the universal magnet, can provide the best metals available, which can then be put to the best use.

In addition to the above, we must also add the following information which should prove useful to you and it is not at all astonishing that both the refined minerals and the crude minerals can be used in increased amounts after contact with our Philosophers' Stone, so that even the volatilized sulfur can be fixed and then volatilized. Whether it follows that by such an operation (which must not be attributed, at least in part, to the universal magnet), the Spirit of the World resides therein and participates in the same, when a mineral containing sulfur will then be in a crude state, as indeed it almost always will be, it should be taken, registered, and subsequently subjected to calcinations by our apparatus and then nitrified. This material is, in turn, pulverized and in this condition fused with gold dust, along with some of our Philosophers' Stone, whereupon in two hours it will be turned into a perfect tincture of silver, where one part in a thousand is extracted in the gold dust, whereupon the Secret of Secrets will be revealed.

Not a single one of all the great secrets will be revealed without knowledge of our magic and the Philosophers' Stone, and I tell you that the great Secret of al Secrets, which from the beginning to end reveals all things from Almighty God, established by His holiest principles that which also abodes forever and will be governed by Hi, so that all things which you discover will be dependent on His revelation to you. So, mark it well and fix it foremost in your mind that the work of God's only Spirit enables you to participate in all of this, so that you will be revealed to you will be brought to light by the Eternal One.

We will now further pursue our revelation in an exemplary manner, and with God's Holy Spirit make a proper introduction to the truth of our knowledge of philosophy in a worthy manner and point out how everything flows out from the Eternal Center and enlightens to His Holy Nature, but before we write about universal truths applying to both the world of the microcosm and the world of the macrocosm, we will start with the realm of minerals as the beginning in which imperfect and impaired metals are found in the brilliance of the Sun in all its glory, and this provides very little for us to use.

Now, then, take in the Holy Name of God, the crude gold-bearing mineral cobalt, discovered and obtained in the mines, as previously noted, was used in the secret preparation, which was then placed on a flat piece of tile and mixed with a little Nitre and allowed to remain for 2, 3, or even 4 hours, whereupon an astonishing change will be noted in the manner in which this material behaves, since there is now a penetrating force of the General Spirit of the World; which shows an increased oscillation and finally, to a very pronounced degree, reveals its astral nature: on heating to a brilliant red, an artificial ruby-red gemstone of inestimable value, exceeding the value of gold and termed the Treasure of Treasures results. When, however, the ferment involving an artificial gemstone such as just described is set aside you have a tincture suitable for preparing all metals.

When, however, you observe flakes of gold in the artificial gemstones which have been heated in our astral natural fire, you will observe that after vitrification, it becomes a royal treasure, and such a magnificent stone of this nature has never at any time been obtained in the natural state, regardless of the length of the search. And this is still the case at this present time.

How, and in what manner, can a gemstone be produced in an hour?

Take a quintel of fine gold, put it on a flat piece of tile or flat porous rock in front of our apparatus and calcine it on our astral refractory hearth for a quarter of an hour, then allow the entire focus to fall thereon and add to the gold a little nitre whereupon it will melt and become a deep ruby-red glass, which will be suppurated and will increase by about a half again. Then add thereto additional pulverized nitre, and then extract the product with alcohol and concentrate the extract and the fused material and metallize the same in the usual manner, converting it into a glassy material by heating (vitrification) so that this gold-containing glass-like material will be three times the amount of the gold originally added, as we weighed before and allowed to flow together from our refractory hearth, will be a homogeneous deep-red liquid, which will cool to a brittle mass and this will be one of the particular tinctures. This will now be one of the thousand parts of silver that has been converted into fine gold.

In a similar process as occurs with gold, so also it takes place with all other metals, and the same can occur in the same way with a glass-like material to make the tincture as with the gold, depending, however, on its quality and properties. Even as iron is itself a metal like all other metals, no yellow tinge (gold) should be expected with it or with any other ordinary metal.

As long as each one produces only its own kind and in no way departs from its own nature, it is necessary then in all things ordained by God for men to possess by virtue of their nature, without understanding everything, for men to make corrections or changes in all things.

But now, to come to the most universal case of all, and to obtain for ourselves by God's permissive grace this greatest of all knowledge (science) and then also to seek to make it, at least in part, worthy of the greatest value, or to make it common knowledge, and this will lead to a very great amount of reflection and thought, and will awaken our understanding, however great our expectations and our confidence in ourselves, that such revelations (evidence) will not be ridiculed nor kept secret, but will have the necessary and most widespread circulation and propagation. Seize, therefore, this great opportunity and take advantage of it, in the Name of all that is holy, the Three-in-One, the true and eternal God, and write and say, so that it may become common knowledge, that even to those very great, though often distorted mysteries, nothing further will be added, and then, with the permissive will and grace of God to make changes in material that has been in existence from the beginning of time and still is no different from the eternal creation of God, through the Holy Spirit, which provided this dwelling place there in the elements of the air, and has given such peace as now you know and knowledge that such a dwelling place of the heavens is the home of the Universal Being, the spirit of our magic we are able to develop in a natural way, thanks to our common love of God, which lets so many celestial things be known to you sinful men.

Now all of this calls into account your loyalty and sends you out into the world with good intentions. Taking our celestial power, i.e., by using the Niter, placed in a wide shallow glass container and set in front of our apparatus containing the magical Fire, and allowing the material to be calcined by the rays of astral radiation, so that the universal, solar essence will be carefully absorbed by its magnetic force and it will become very finely powdered and will increase in volume and when it has been sufficiently treated in this fashion, it will be further treated with alcohol, then heated from time to time with salt, and finally subjected to the full focus of our natural radiation falling upon it, so that it will melt and be purified by the bright sunshine and become absolutely clear, calm and motionless. Also, in the same manner, through the great Glory of God and His Holy Will the Philosophers' Stone has allowed you to launch out on this venture and to bring it to an end. Then you will know for a certainty that the marvelous works of Nature owe more to the Will of God than to our expertise and I hope that you will make it very clear and very evident that God Himself brought this about by providing His support for you. Not in this alone, by giving you such a Stone, but also by laying open before you everything above and below the earth, as well as all the Secrets of Nature, the blessings of Heaven especially have been bestowed upon you, and always by the way of certain conspicuous and eternal truths of God which rival the brilliance of the Sun and of all the stars, continually and endlessly, in those smaller objects which are observed to move and have their being, and which for you lie in the bosom of God, who has provided all such wonders in Nature, even those that are all still hidden from your view and thus are still secret since unrevealed to you. And they will delight you and make you rejoice abundantly on so many future occasions, as will all other men of the civilized world when they learn that they too will have the power to accomplish the same. Know also that God will reveal all this to you with such a spirit that you will be able to observe how all-powerful He is, yes --- and forever even greater than that --- so that in His omnipotence and omniscience you will see that God surpasses all earthly knowledge and wisdom. Since you then are increasingly aware of you dependence on the things of God even for this transitory life and you soon maybe taken to be with God, which you strongly desire and at which time your joy will be complete. All these and much more are the impressions derived from the Philosophers' Stone (The Stone of Knowledge), and all the best and most beloved events that befall you are like gold. Those people in humble circumstances will not forfeit a holy life nor have to give up that which appears to be pure gold, whereby a universal tincture can then be made and can provide a force able to penetrate all metals and as a result can be responsible for making the best gold. Moreover, God has also provided this Stone and many other blessings like resourcefulness, astuteness, and long life, along with additional rewards which have been described in many places, and give for the enlightenment of all. Thus, one can say, in all truth, that it will be all in all, and all in God alone. So be it.

We will at this time also point out that the content of Potable Gold in water is as high as to be obtained anywhere.

Lay a quintel of gold on a flat piece of tile, and melt the gold by our natural process: Fire, and take note! Soon it will begin to flow in liquid form. Then remove the apparatus, so that there will be approximately 2 or 3 less times the heat; the support material will now always involve a small nucleus of suppurated ordinary sulphur, which after calcinations becomes gold, whereupon it will then turn into a bright red color and will be very hard, brittle, and friable. This can quite easily be reduced to a fine powder simply by crumbling with the fingers. This powder is then weighed and will be found to show a sizeable increase in weight, due in part to the fact that it has gained in attraction of the Spirit of the World. Part of it, however, does come from the added weight of the sulphur which has been added and must, therefore, be included. Now put this calcined gold in a clean glass dish and cover it with a suitable amount of good wine or brandy, so that it will dissolve in an hour to give a blood-red solution, if accompanied by vigorous stirring and then set it aside until nothing further takes place. Then pour it on a flask and distill it over, which is easily carried out by our" Magical Machine" [magnifying lens] with "reverse distillation fire", which causes the material to be distilled over to the last drop and will no longer remain in the form in which it had been. This is then the potable gold which has been prepared in dessicated form and is the spirit of pure gold, having a deep red color, a characteristic odor, and a very characteristic sweet taste. For the young, the dosage is a single drop, for older people, it is 2 drops taken early in the morning in strong wine, with a resulting sweating to be expected. In this way you will always be able to cure all the internal ailments that are so detrimental to health.

From the beginning, we have introduced those substances we have taken from the mineral kingdom and proven to be of universal value, and we have already enlightened everybody sufficiently, but now however, we have added other substances from the vegetable kingdom to those found in all noble and base metals, and also all things which live in water (all of which are noble and of value), and even less than perfect pearls and mother of pearl, as well as corals, mussels and various active and inert substances, whether crude or elegant, whether crystalline or amorphous (translucent or opaque), such as now are available and might possibly have been selected, all of which belong to the mineral kingdom, or are dependent upon it, and for that reason are used as examples. We will consequently point out as well how even such can be made to be more perfect and more useful by using our "magical natural fire" and our natural apparatus, as demonstrated. What is now of greater use to us, we acknowledge has been given to us by the omnipotent hand of God and not alone for the transient kingdoms of this world, but also in the infinitely longer life to come in the Kingdom of Heaven and to enjoy the use of such up until the time we draw our last breath and go to be with God. We unfortunately are given His greatest favor and therewith our souls receive forgiveness from His Holy Hand and can be commended to His care.

Just as we, through the miracle of faith in God and reliance on His love, are able to accomplish many wonderful things, and are able to overcome all obstacles, so, too, are we able, through the Holy permissive Will of God, perform our magic and prepare our natural fire by means of the apparatus that we have just now introduced to this world in order to conquer it by the highest degree of completeness and perfection which we are able to accomplish, which would have been impossible without our having received the idea of the "natural fire" to help us. Especially since the value of such a natural phenomenon could require a thousand or more years to be adequately demonstrated and further developed, had its worth not been displayed through the knowledge of our apparatus.

Recall now, and cite it yourself, whether you or anyone else has at anytime heard or seen how ice, talc, pearls and plumed objects (especially since a very strong "materialistic fire" can always be produced) can be brought into a tree flow or to be melted, without putting over a fire. As you should know --- and no words of praise are necessary here --- that it is reasonable to have this property, that is, to possess our "natural fire", apart from which all this would be impossible, as we will show by subsequent demonstrations. We will now take, as an example to prove a point, a few valuable oriental pearls and place them in an open glass dish in front of our apparatus, so that the pearls will lie in the "magic natural fire", with its focus at the center of the dish. The pearls will be irradiated thereby and the effect will instantaneously be indeed marvelous and you will see how nature in such a short period of time can bring about the calcinations of the material with complete fusion, accompanied by the loss of water, so that there is a sizeable increase in volume of material which will itself become more valuable and move, by degrees, into a higher and more permanent state as you will soon see.

At this point, you should know that these pearls were obtained outside their natural environment and were also fixed permanently by vitrification and produced by bringing many inconceivable celestial bright-shining colors to those magnificent gemstones.

Oh Glorious God! How great and merciful You are and how you manifest yourself in so many ways, even to us who are so sinful. Forgive us, foolish and unworthy creatures that we are, and show Thy great mercy to all of us and Thy great goodness to all such as we, according to Thy great glory, so that we may love You and praise You. Call us into Thy service, so that we may please Thee, for we were created to be Thine.

Because of the great honor that belongs o You, which You so richly deserve, since You are our Master, even as we are Your servants, which as You have indeed foreseen and do know, that Your love will serve most of all to protect us and be the secure foundation of all that we hold dear, so that even for many thousands of years hereafter, the entire host of obedient souls shall with endless compassion and hope then be able to carry out your will, do good to others, seek your blessings in all things, and bewail our transgressions of Your law, for You are our Lord God, even in this transitory life, which you may choose to prolong by Your undergirding support.

Now this noblest and altogether beautiful and magnificent stone is that which serves as an ornament to all the kings of this world and is the ornament of preference because of its greater beauty and indeed, it is priceless, since its value exceeds all others beyond measure.

It is the same stone which God in His excellent
goodness and glory sees fit to grant long life.

It is the same stone which is able to cure every day
whatever illnesses and sorrows there might be.

It is the same stone which can increase the
gratitude of mankind and take away all manner of
evil.

It is the stone which can make men feel younger and
desire to produce new fruit.

It is the same which can, through the Providential
Will of God, bring about life where there was death.

It is the same which enable man to walk with angels
and converse with spirits.

Yes, indeed, it is even a manifestation of the
secret mysteries of the whole world.

Oh Beautiful Stone! Oh work of angels! Oh,
everything good and perfect, for which we thank our
God and are eternally bound to Him.

We will, at this point, allow you to indicate how you should value this stone, given in such honor and so useful to God, and how you will be able to put it to use yourself.

In the first place, you should give the Lord God all the honor and glory and love Hi who has protected you, and rewarded you with gold, although not merely because of the great love that He has shown for you, so that you will feel safe and secure and free from the devil and from all kinds of spectral apparitions, and will be totally dependent upon Him. He will help you to recover from all temporary reverses that occur such as those which take place in times of pestilence, and (like an amulet) He will protect you completely. Regardless of whether or not such things have already befallen you in this life, you cannot be overcome by them and you will again follow all safety precautions. I say to you in truth herewith that this stone which has been revealed is indeed of universal value.

It follows accordingly that otherwise non-flammable aluminum will be covered with a feather-like powder [aluminum oxide]. Where you do not also obtain such in crude form from the mountains, as this jewel contains our secrets, and yet everybody has evidence of this "natural fire", in just such a glass dish and wit our apparatus, it can be attempted in such a way that you will certainly have a very noble gemstone. And if you will look about, so you will note to what extent the aluminum is still covered with the powder (which has a purpose) and what this gemstone will finally turn out to be (for indeed it was formed increasingly by our "natural fire" with many colored flames and has been given into your powder and you will be responsible for it).

It may indeed prove to be that which came from our true electrum, from which, as was earlier reported, the basic tincture for the improvement of all ordinary and all secret stones comes, wherewith to bring it into perfection and to its highest quality, which since our universal stone must offer a hand and provide help, and since the magnitude of this fire is found to be small, although in itself it has been able to fix noble material, as indicated, then you will see a fundamental essence that is inaccessible in a materialistic fire! Yes indeed, you will see it --- and we say to you to examine yourself closely for this nation [Bohemia], in which we are now writing on this subject, contains one of the greatest secrets, about which royal secrets more will follow.

The behavior of the talc in the fire is described quite well. Like true artists, we let everyone come before us who can praise the quality of our talc after surviving the "materialistic fire" and completely recovering. We will point our however, that apart from its astral nature, there can be no fire of this type. Moreover, this has never happened before. However, it is our wish to, and we herewith will do so contradict all sophisticated interpretations, therefore, and prove by means of our apparatus and natural fire that the talc, like the aluminum powder can be melted and poured off after a very short period of time and be converted into a tincture, as has been demonstrated earlier. This tincture does not undergo further appreciable change, and is much like its predecessor, in that a common ordinary stain can also be added to it in more or less considerable amounts. We will now begin to show how this can be accomplished. This is done by making significant tests on the talc oil and noting what its uses are.

Nor is our true "talc oil" alone in its usefulness. There are also other oils which come from metals, minerals, pearls and gemstones which can be, should be, and must be brought to completion through no other art than "our magical fire", which must be used with considerable emphasis on safety, by carrying out the reaction in which ordinary mountain talc is put in the aforementioned glass dish in front of our apparatus, where there must also be a calcining fire, so that the material can be calcined until a blue powder is obtained without any additional smoke, after which the apparatus gradually takes up the calcined material after only approximately 2 hours of digesting over the fire, so it will come to pass and all the world will regard it with astonishment to see how strongly the mystery of this calcined material with its own ascending spirit frequently draws the spirit of the world to itself, resulting in the formation of a highly colored oil and a gaseous product. However, except on this basis, while its role will be maintained, which will establish it, and which will influence the entire world are questions to be answered. Mark this well! The more a metal or mineral is calcined by our magical "natural fire", the more it will go into solution. The more often it subsequently draws off the spirit of the world to itself and dissolves to produce an oil and a gaseous product on heating in a strong fire (we know our fire very well), the more it becomes coagulated and the greater the degree of fixation (and vitrification), and the better the results obtained for the health of men, metals and ordinary stones, for which God merits eternal praise.

We will also now subject the vegetable kingdom to our magic of resolution by use of "natural fire" and carry out tests by irradiating briefly, without relying on fire of an astral nature in anything at all, and to cause that which lies in all nature, as established from the beginning by God, as he ordained.

Thus, it is a great pity that so many thousands of men, with so much money and this world's goods, are plagued day and night, for no purpose at all, and weary themselves in what they endeavor to accomplish with their magnificent invention of ways of distributing their wealth --- ways which depend on their materialistic fire by digestion, calcinations, distillation and many other methods of heating in order to bring in a good return on investments, according to presumption. However, your wailing is without end, and were such not maintained, we would be able to demonstrate this definitely with our "natural fire" (which is entirely and completely natural) and can produce light in a very brief period of time, for our astral process of natural fire can produce, maintain, increase and bring about everything which, on the other hand, impairs, reduces or totally destroys your "materialistic fire". As a result it can clearly be seen that little, or even nothing at all, that is good can result. On the contrary, however, our naturally occurring fire, which is nothing else but the blessed sunshine which leads to the bringing forth of all things in thousands of different patterns.

We will not consider by what means and in what manner the five essences and tinctures of essences can be made and preserved from all vegetable matter and provide from the very beginning the most valuable white particles.

Here it is to be noted that there are two kinds of tinctures obtained from vegetables. One gives a greater amount of the same, while the other will result in producing better results from the standpoint of health, and should on the basis of merit and increasing tincture production become rather cheap and so preferred by us from the start.

In the same manner as we have described in the most universal way, so we have discovered its nature and qualities.

Thus, when you take coagulated spiritus mundi (so-called) and calcine the same without natural fire (so designated) up until the same turns yellow and continues to remain yellow and then turns red, and then remove it according to the type of apparatus you have available, and the size of the magical natural fire, as well, so that it will be only slightly digested and thus will attract the spirit of the world with incredible surprise as its very next blood relative in such a manner that it is strong of itself, so that in a few short hours it can serve to measure the whole blessed world and reflect its spirit.

When you wish to note it down here, note it carefully, and where you do not, it is not given from God for you to do so, then we can do nothing further about that here than what we have said in these few remarks, in which all things and mysteries are concluded and laid to rest, on which in the final analysis the entire concept depends. Thus, we should so arrange our previously disorganized world. The spirit thereof is somewhat closer to our magical fire by subsequently thickening, by boiling down, little by little, until the result obtained is a multi-colored powder (if I may be permitted to make recommendations involving the three characteristics, then I would suggest the stone which seems whitest to the eye) which is presently the tincture prepared from vegetable matter and, in what is to follow after this, you will find the same should be used.

If you will now put a small part of the above powder in a large amount of fresh water, then it will immediately begin to increase in amount, so that the thousandfold increase will be unheard of and never seen before as the greatest comfort given by a Holy Blessing in the same manner in which the white particles produce an increase and this also can happen with all other kinds of scattered seeds. However, all that has been grown by planting, cultivating and allowing to reach maturity, and such growing results in an increase in the things grown. Furthermore, the roots should be covered over by this colored water.

Now, it is known in every case that God, out of His love for men, has created all things for man's benefit and such as are created for such use and adapted to man's need will serve as an example of how all vegetables must be treated, through our "natural fire" and all distillations proceed through our apparatus, as conceived to take place from the start in our magical natural fire, which you will understand clearly from the tabulated information and the glass apparatus that we build.

In the first place, the glass was filled with rose petals or other flower petals, heaped up, and ten made ready for the reaction by making it or loading it at the top and placing it in a ring on a mold or on a table in front of the apparatus and let the "natural fire" do its work without reacting too strongly thereon, so that carrying out such an operation first involves the idea, and then the fact that the oil level drops in the glass or the spirits are eliminated. This is a thousand times more efficacious than when such is expressed through the "materialistic fire" and the ordinary art of the apothecary. If it is your wish, moreover, to lay out in further detail a still greater secret than the above distillation operation, then it should be set down here and now. It should include information on how tinctures can be made from flowers and other plants.

If you desire only to make tinctures of increasing amounts of plants, then lay them in front of the descending spirit together with its oil in an open glass dish and allow our magical fire gradually to irradiate the apparatus containing the sample. You will then observe with considerable pleasure, how these flowers produce volatile materials along with their oils by means of which the spirit of this world (i.e., alcohol) extracts from it colors which such flowers contain or even developed on its own as a final product, after evaporation, and then becomes a very mysterious secret, which thousands of serious students of the art cannot understand. Moreover, through God's Holy Will and the wonder and praise which this inspires, by the addition of a small amount of this tinctured powder to ordinary fresh water, so this material will be tinctured to the best rose color or other colors, which can be prepared from white flowers using distilled water, as reported earlier. Subsequently, a Royal Treasure, held in high regard as the "White Solomon Seal" has been obtained.

Should we place our reliance on that, and arrive at a universal value resulting from the consequent selection of the best wine made from the best grapes and describe tinctures of the same, with wide usage as might well have been expected, in small doses, then many hundreds of types of wines can be purchased and sold.

A single wine, however, drunk only occasionally at most and then for health's sake and quite agreeable to the body, is poured into a shallow glass open dish until full and then placed before the apparatus, so that the rays of our "natural fire" will be focused without obstruction on the wine in such a manner that it will gradually become hot enough to boil. However, do not let it boil, but gradually reduce the heat by about half. Then, add more wine gradually over a period of time until 3 or 4 or even more volumes of the hot tincture result. Then, increase the amount of heat and let the mixture continue heating an hour longer, still keeping it liquid, then increase the heat by about one third more, at which time the spirit of the world (that is, the alcohol) will be reduced and coagulation will set in. Then push the apparatus aside and let it cool down. You will then have formed a gemstone which will have the power to change water into wine. In order to give this tincture a higher content of alcohol, you must place this tinctured gemstone in an open glass vessel in front of our apparatus, using a very gentle natural fire to dissolve it and then in a short period of time it will become very fluid, at which point the stone will be decomposed completely and be calcined by our fire, so that the alcohol will be reduced to about half its original volume and by increasing the concentration of the other components in the mixture will become far more fluid so that now, as the result of the smallest amount of heat from the sun, it becomes dissolved in the liquor, which acts more vigorously the closer it is to our "natural fire", until gradually heating more and more it will finally have been converted into a gemstone.

The amount of tincture obtained by using this stone is 10 times greater than that possible when the stone that was prepared in the first instance was used. Moreover, using this tincture it is possible to turn water into wine, as well as always to improve and lengthen the lifespan of all mankind and in this way provide a most useful service, since the more frequently such improvements in the lifespan are repeated, the greater the increase in riches will become, and for this we should be in the eternal debt of our gracious Heavenly Father.

Afterwards, we will search all through the mineral and vegetable kingdoms and then to some extent, after sufficient praise to God, we also observe wonderful things in the animal kingdom and in carrying out our research in view of he many possibilities shown by Nature and so many the number of choices permitted us by God, which we have mentioned herein and have described to you, we will assert the following:

Let him be accursed who misuses the greatest gift of God, who has shown us grace and mercy, and instead resorts to a diabolical evil.

Woe unto him also who God has given occasion and permission to write down heavenly things and who is known to give unworthy stewardship to this opportunity.

Let him be further accursed and condemned who uses our writings and holy wisdom in a special way to further his own interests or to bring to himself or others anything other than their use to the Greatest Glory of God.

Let him also be accursed and damned who uses our writings in any way that is false, mean or contemptible.

All of you mark well that such curse will not remain exclusively on you. While it concerns itself with the highest majesty of God Himself and thus with His Holy Omnipotence, there is nothing that is prescribed in our art and science that is, outside the permissive Will of God and to which He has not hitherto given His blessing, so that all things that we do have been carried out with the idea of usefulness and tranquility in mind.

All of you who thought previously to restrict yourselves to the animal kingdom should know at this time that in that kingdom it is of the least importance to discover mankind for ourselves and to be unduly influenced thereby.

So set herewith an example and examine yourself to see whether you will be given over to an animalistic urge in a philosophy along with the blessing of the wisdom anticipated to result from our magical natural fire proving to be of value only to a limited degree and for a short time, which otherwise might turn out to be of inestimable value to you.

Nothing that we can say to you, except to tell you about an imperfect though very aggressive creature, whom we shall call Homunculus who is represented by a small figure involving only a few brutish creatures, originating outside of nature and contrary to the Will of God and coming from the Devil himself, incarnate in such a profane form and appearing to all mankind as a ministering spirit of the time, and spreading everywhere around the idea that he is concerned with the souls of human beings. For this reason, we concern ourselves in this, and on the basis of descriptions of many different homunculus types, we will open our minds to God's permissive Will and dwell upon useful thoughts and ascertain what good things in Nature are to be further achieved.

Since time immemorial nature has gone by God's providential arrangement, which includes provisions for you, but through His permissive Will, and help from our Art and Science, it is possible to shorten the time and occasion the birth of ideas and to demonstrate them and subsequently document them.

By taking many individual hens, ducks, geese and other poultry and birds, taken together as a whole, in a cage or nest made of flax or wool, and placed in front of our apparatus under our "magical natural fire", as close as possible, then the cage or nest will not be set on fire, even after 3 or 4 hours, although this might not be the case after 4 weeks.

Many wonders are still to be found in the flesh and bones of mankind (all of which have been described to some extent), and in his excrement and urine, and it might well be possible, by means of our "natural fire", to be able to come to a preliminary picture thereby of very important, yet presently very obscure, things and predictions about them as a result of the death of certain living creatures, even though the evil of men and the tensions resulting therefrom might greatly displease God. Let our thinking processes remain on this matter and we will continue to study it.

It is possible thus, to have good character and at the same time perform wonders, even miraculous, deeds.

No one who is truly blessed likes to boast about his good deeds, or set up something sensational about himself or to introduce to the world some special wonderful thing that he has accomplished, still he has, however, attained something worthwhile by using our magical fire, of necessity however, in conjunction with the position of the appropriate planets in their course. Since, however, it is not uncommon that such events can happen, we will not place briefly --- but reliably --- before your eyes and give you full information about a Magic Mirror, so that you may be able to determine, among other things, the initial time in which the weight of all 7 metals can be obtained simultaneously so that the purest one can be selected, and then one after another selected on the same basis, as well as on the strength of their reactivity at that time, determined by melting in an appropriate glass crucible at the proper hour of the planetary sign in front of our magical natural fire.

It is also true that the gold must be melted on a Sunday and under its own planetary sign, since the sun itself is considered a reagent, just as objects are melted when focused in front of our "natural fire". This is followed by Silver on Monday, since the moon is likewise the reagent for silver. On Tuesday the same operation is carried out for iron. On Wednesday, it is with mercury, Thursday with tin, Friday with copper and Saturday with lead.

This is the manner in which the melting process occurs and so each one of the metals will be found to follow a weighing sequence which indicates that there is a fixed pattern that is constant and can at all times be subjected to testing, which reduces the upper and lower members of the series by the same appropriate amount, while the force involved in every case is the same, even when the series is rearranged. Thus, in all these signs of the sun and of molten objects, whose characteristics show that they are the result of natural forces and at the same time the result of supernatural forces, both of which affect their motion (behavior).

Consequently, all 7 of these purified metals act in concert on a Sunday just before sunrise, making possible a mixture of metals and the pouring of them into a previously prepared mold to give a mirror which can be polished. The master craftsman responsible for polishing the mirror at no time examines it with the naked eye, but at all times wears special eyeglasses, which provide a sound basis for turning out products of assured value, and charge him alone as the guarantor of it serviceability. The same craftsmen from the very first moment visually inspects the mirror, inasmuch as he is a master in his field and this lies in his area of expertise and everything that is asked of him and everything that is required of the mirror is, naturally shown to be valid for such a mirror, when he is demonstrating it.

From this apparatus a wonderful magnetic bell can be cast, by which an entire army might be summoned or be sent into retreat, and as a result, it should be rung or tolerated in such a manner that its frightening tone requires considerable force to produce it. From among the 7 metallic signs, we will call attention to the following:

That of all these, or even any one metal by itself with its own set of properties, can be made by our own apparatus and natural fire and must be cast at the hour of its formation, is both foolish and invalid. Idle talk about testing must be terminated and must remain our best kept secret, so that we can retain it through constant diligence in observance of God's Will.

Moreover, in the end, at which time we will complete
this life and return our spirits to God who gave
them to us in the beginning so that we might know
His eternal and blessed state of peace and be happy
to have attained to the same.

We could elaborate further in concluding this work
and mention briefly, so that only one or two men, or
at most three, in any year could consume his usual
diet, using our natural fire and our apparatus
(despite the fact that the container might have to
have a cover to that the radiation might not fall
thereon) to heat the material and then enjoy it, and
consequently obtain a long and rejuvenated life, as
well as providing, with God's help, a happy ending
to this life and the assurance of entering Heaven in
the eternal life that is to follow.

Amen.

                    Finis.

Solinus Salztal

Discourse on the Fountain of Philosophical Salt

Translated from the Latin by Patricia Tahil

I happened one day to be taking a rest from my work. I was not thinking about chemistry at all. So I began to put my laboratory in order, picking up the glass instruments, pots, and other vessels that were lying here and there. I also repaired some furnaces. Finally, when I had spent the whole day on these tasks, I was tired, and sat down on one of the chairs at the table. Sleep came over me very swiftly. I had scarcely closed my eyes when someone opened the door. I saw a funny little man. He greeted me warmly. And said that he was a student of chemistry and wanted me to teach him. He said that he had come to meet me because he had heard about my work. I began a conversation with him and asked him how he liked my instruments. Then he asked me which operations I used them for. I told him that I was looking for the stone. He smiled and said that he thought I must be using such remarkable instruments to deceive careless people. He said I would see that I had really been deceiving myself instead. I have never let embarrassment get in the way of my learning something. Besides, I could see that what he said was true, and I was well aware of my shortcomings. I asked him whether he had an easier way, and whether I neede more instruments of various kinds. He said, "I see that you don't consider it a disgrace to learn something and to admit your ignorance. Many people I have visited do feel disgraced. But you want to profit from what I have to say. So I will show you things that very few people out of many thousands have ever seen". When he had said that he started to leave, and I followed him. He looked back, saw me following him, and said: "Now I know that you have a great desire to learn, since you are determined to follow me". Then he took my hand and led me out past the city gate. There, put in to shore, was a small boat called Reason. We got in and set out over the water with the help of two oars. Soon we could no longer see the city, which was called Ignorance. We had just passed by the towers of a city called Arrogance when we caught sight of a high mountain, it could be. He told me it was the Salt Mountain, and that its salt water gave moisture to a kingdom located in that region, a kingdom called Earth. The water made it so fertile that animals, plants, and metals grew there in a most remarkable way. And if that mountain were not there, the entire fabulous kingdom would perish in an instant. But as long as the mountain lasted, the kingdom would be so abundantly fertile that it would never lack for anything it needed. To promote this great fertility, gold grew there in such abundance that there was always work for the miners. When I had heard this, I answered: "If this water nurtures all things with its power and consequently causes even gold to grow, it should rightly be called the moist radical, or philosophic mercury". He said: "I see that you have spoken intelligently. You have observed that this water of the wise is quicksilver, the same thing that makes metals become alive and begin to grow. So that you can learn more about this matter, we will direct the course of our boat to it". I obeyed his words and took up the oars with a sorrowful heart. With words and labor I tried to move the boat forward, and in a short time we had made good progress. Finally, after great effort, I put to shore at the salt mountain. Then I anchored the boat called Reason so firmly that the waves could never move it from that spot. We began to climb the mountain, which was dripping with moisture, and after twenty paces caught sight of some hermitages. Here members of the Rosicrucian Order were living on the fruit of the mountain. The old man listened to what they were saying, and then led me up to the mountain. I saw a sumptuous fountain springing out of a statue of Venus. Its water was salty. There was a stone basin to receive it, and in the middle of the basin stood the statue, atop a white swan. The water was sent down through various small pipes into that stone coffer. Some of it overflowed the rims and ran off. Along the fours ides of this fountain stood four animals --- a green lion, a white unicorn, a basilisk, and a dragon. On their backs they bore four white marble columns, and these were joined at the top by arches in the form of a cross. Mercury was sitting above the crossing, and on his head a winged Fortune was standing on one foot. As for the nature of the water, it stayed salty for a while, and was shimmering in color, though really transparent, clear, and crystalline. Then, because my servant wanted to have something to remember this moment forever, I told him to climb to the rim of the fountain and draw up some water. But he leaned over too far and fell into the water, and before I could get hold of him, he vanished under the water. Part of him was transmuted into water and part went up into the air like smoke. When the old man saw that, he handed me an optical tube so that I could watch where the vapor went. I saw it filter down to a certain traveler walking along the road. He took it in with the air he was breathing. Then the old man said: "Behold, now the water of the fountain is prepared. Its water has been transmuted into animal substance. The water from your servant has brought that about. Now it is like female seed, ready to receive forms from animal sperm, if only they can come together". As I walked around the fountain and looked at it closely, I saw a pipe coming down from the mountain, a pipe that was connected to the fountain and had a stopper that was easy to open. I asked the old man what it was for, and he told me that the fountain flowed away from that place. Moistened the earth, and united to it all kinds of animal seed. Then all kinds of birds, worms, and animals grew from the union. But when the stopper closed, the fountain could not flow out. After this explanation, the old man went on to say: "Behold how powerful this fountain is. It joins itself with animal seed and grows along with it. Do you see how it frees all animal bodies through its own ardent love, which is charged with magnetic force? It makes whatever is fixed similar to itself; but whatever is volatile, namely, masculine seed, it leaves free to soar. When the seed comes back, now grown heavy, it fixed it, so that the fountain can carry on the solution and coagulation without pause. Therefore, if you can purge dissolved water, recombine it with fixed water, and then fix it, you will have a much more excellent compound than you had before. Using this water, you will soon manage to get a deeper understanding of all animals. You will then come up with the true philosophic medicine derived from all animals". Then I asked the old man the name of this water. He said it was called universal doubled mercury, also microcosmic mercury. Then he said: "The water of the fountain was once universal at all three mountains, but now because of your servant it has become microcosmic. So an inscription has been placed over the fountain: Make mercury through mercury through mercurial water. The smoke that rose up from the fountain your servant fell in is called sublimated mercury. It was bound and constricted by means of fixed water, and now by the same means it can be bound over again". Next I asked him the location of the fountain, and whether there was any other like it. The old man replied as follows: "There is no other fountain like it among all animal things. Only this one can bind itself with the volatile, sublimated mercury derived fro all animal things, and only this one can put on their form. The place where it is found is called Pansoma". Then the old man ordered me to fill a phial with that water. When I had filled it, I tried hard to glimpse my servant in it, but he had dissolved in the mysterious, greenish water. Afterward we left the first fountain and moved on to the second, which was sumptuously decorated wit a statue of Venus just like the other one. Its four sections were connected by four elm trees forming a cross from above. Alongside were vines with clusters of grapes on which another Mercury was displayed. As

before, Fortune stood above it. The properties of
this water are the same as the water of the first
fountain before it was changed by my servant. Then I
asked the old man another question about this
fountain --- if its water were sprinkled over the
earth, would it cause it to grow? He said it would
not: "this fountain cannot unite with any sublimated
mercury unless it has been fermented with the fixed
salt water of its kingdom and transmuted". Then the
old man took a knife and cut off a cluster of grapes
and threw it into the water, where it vanished at
once. A fine vapor rose from it. At that moment the
old man handed me a phial in which to catch the
vapor. When I had caught it, he poured in a little
of the salt water from the fountain. In the middle
of the phial a volatile vapor, once again fixed,
took form. A stone also took form, which he called
the vegetable stone of the philosophers. He said
that in this way all the best essences could be
extracted from vegetable things, if they were
dissolved in this water and again coagulated. When I
asked him to open the small pipe leading from the
fountain, he complied at once. With the pipe open,
all the earth around it was moistened with dew, and
in an instant everything sprouted, since that water
had been joined with all vegetable seeds. Then he
looked about for a bit of silver, and when he had
thrown it into the fountain it grew up again in the

form of a tree. Behold", he said, "how this fountain itself becomes vegetable through the vegetable fermentation of grapes. So it has also drawn out a metal into vegetable form". From this I inferred: "Therefore an animal fountain will draw out things from the other kingdoms into animal form". But he replied: "You are asking too much of me; be content with what I have already said". And so he closed the pipe and the fountain ceased to flow. I took the phial filled with water, which was of two colors, white and green, and went with the old man to the third fountain. It was magnificently decorated with a statue of Venus and four columns. One of the columns was made of gold, the second of silver, the third of copper, and the fourth of tin. They came together in an arch at the top, and a sheet of lead was paced over it. On top of this sat Mercury with Fortune, arranged as before. When we drew near this fountain, the old man addressed me in these words: "Behold, at the other two fountains you saw marvelous things that you had never seen before. This fountain contains the explanation of the two previous ones. Here is the foundation of all hermetic knowledge. You will take it in as if it were formal instruction. Therefore, pay close attention to what will be dealt with here". Then I repeatedly asked him to begin the instruction right away and to let the fountain flow free, so that I could see exactly how the mechanism worked. The old man told me it was impossible to do that. It would require an order to make the fountain metallic. And it would have to be done by means of salt water that was already metallic. This proves the truth of the words written around the fountain: Make Mercury through mercury through mercurial water. I asked him how, in that case, I could obtain metallic mercury. But he said: "Do you see what material the fountain is made of?". It was made of grey stone with many veins in it. "Don't you see how many fragments the inhabitants of this mountain have chipped off? Surely they must have done so for a reason". Saying this, he gave me a hammer, and I pounded off a piece that weighed the same as half the water in the fountain. I threw it into the water and it vanished there, while the water stayed clear and beautiful. But it lost its brilliant gleam. As I noticed this, I saw waves like surf suddenly stirring in the fountain. But they gradually diminished, and most of them grew very slight until they turned to black ice and the fountain dried up completely. The old man saw it and said: "Now the union of universal mercury and mineral mercury is complete. Now the transmutation of universal mercury to mineral mercury will take place, and the manifestation of mineral mercury by means of the universal --- a task that took only three minutes". We let the operation rest in that state for half an hour, in order to make sure of it. By then the fountain began to flow again and was as white as snow. At that point the old man said: "Behold, now I have the doubled mercury in my possession. Now I own it --- white lily, powder of adamantine, chief central poison of the dragon, spirit of arsenic, green lion, incombustible spirit of the moon, life and death of all metals, moist radical, universal dissolving nutriment, true menstruum of the philosophers, which without doing any damage or harm reduces metal to first matter. This is the true water for sprinkling, in which the living seeds of metals inhere, and from which other metals can be produced. Through this water their potency remains in solution in this water. In all kinds of aqua fortis and other such unknown philosophic waters, they lose and relinquish this potency. In this exalted water is the true vitriol of the wise, of which Rupicessa said: "Vitriol or salt is the proper seed to generate all metals, including both the remote and the proximate seed". I will show you its power as clearly as in a mirror: for this water from the fountain radically, silently, and wondrously dissolves all metals, white and black, by its own innate power and magnetic force. In an instant it liquefies metals by its own internal fire. It opens their pores and enters them like feminine seed, attracting the masculine sperm to itself as if it were attracting the soul of the metal. It leaves the lifeless body behind like refuse that cannot endure the fire. Certainly it is a very marvelous thing that this water strips metals of their dignity. It is the dry path of the philosophers, by which metals are reduced to their first matter. It is considered very swift, but compendious. Since we want to proceed on the humid path, in which common water is added to this water to make it liquid, we must first make the metals very bright. This operation takes a great deal of time and effort, but it is beautiful to look at. This is the principal operation of philosophic mercury consisting in the radical solution of metals. They are dissolved away from their seeds silently, by a force of burning love. From this principal operation of the water, all the rest follow, beginning with coagulation and generation. This saying of the philosophers applies: "The corruption of one thing is the generation of another". Thirdly, it is called philosophic medicine. These three secondary powers contain countless others within them, all of which arise from the first radical solution. Up to this point, you have perceived the qualities of philosophic mercury as the feminine seed of metals. What follows is everything about the correct use of the philosophic stone". After he had told me this, the old man showed me that I would need to take earth, the matrix of gold, and put it in the water. I asked him where I could get it. He replied that the same substrate that contains mercury also contains the metallic earth of gold. He gave me a hammer, and I chipped off the same amount as before. When it fell into the water, he asked me to give him enough gold to weigh the same as one sixty-fourth of the water. I handed him four Hungarian ducats, which he filed down and planted in the earth of the fountain. Then the water, white from the earth, gradually begins to change color to red and finally dried up completely. The old man continued: "Now the union of gold earth and philosophic mercury is complete, and the principium of the stone has been made. Now follow the radical dissolution of gold, the manifestation of the seed of gold, and the radical conjunction of the principia". This task was finished in half an hour. But the water had taken on a purple color, and when I realized that it was bland and tasteless I cried out in terror and asked what had become of our Mercury. The old man ordered me to pour in some common water and begin the process of extraction. When this was done, the water changed color and became salty --- but not corrosive --- rather than tasteless. The old man ordered me to dry and clarify it, since no more water could be extracted. When this was done, there was some unstable white gold in the dregs, and I saw that it had been robbed of its soul. When it dried at the edge of the fountain, a rainbow appeared in simple form with all its colors, and from it the golden water of the Cabalists proceeded but soon vanished once more. While the fountain was drying up and leaving behind the red dust shot through with redness as intense and vibrant as that of the sun. I took out this dust and put it in a phial. Then I asked the old man what I should do with it, and got this answer: "If you extract this dust with fine, burning water and concentrate it, you will have true potable gold and a philosophers' stone that is not yet altogether fixed and is useful in the cure of all illnesses. But if you coagulate it for a long time and fix it in fire, you will possess a permanent, fixed philosophers' stone with which to cure metals. Taken by itself, this dust is called the first matter of the stone, because all three principia of generation are subtly brought together in it. Here also the seed of gold, meaning its feminine seed, and earth of gold are joined in correct proportion. Therefore, if you can enkindle the natural fire hidden in the seed of gold using external fire, and cause it to look for nourishment which it converts into itself, you will have something to rejoice about. Only make sure that you have been properly instructed, since metal seeds draw in the same amount of saline water

as you need for fixation. If you add too much of
that water to them, they will dissolve before they
can be fixed". Then I asked him what was the name of
this, and he said he called it mercury of the
philosophers. When we had thoroughly looked over the
three fountains, we began to climb back down the
mountain toward level ground. But the old man led me
into a mountain cave where there lay a magnificent
statue armed with a two-edged sword. I asked him why
the statue was there, and got this response: "There
is on this mountain a spring that belongs to the
three fountains. This statue, indeed Nature herself,
guards the spring and keeps the three fountains from
ever going dry. The spring originated in these
fountains. Nevertheless, it was connected with them
for this purpose also: for the spring water to fill
the fountains, called Pansomata, as it went up the
mountain; to take animal, vegetable, and mineral
operations from them, afterward to pour itself over
the seeds. It grew along with the seeds and left its
form on their dead bodies. Then it became spring
water once again, again went up from the spring to
the fountains, and again took on a new form in them.
When it went back to the spring, it lost this form.
And so it was in constant circulation as it went up
and down the mountain. In the spring was first
matter --- formless, omniform, and of single form.
But when it settled in the fountains it became

second matter, known as doubled mercury of the philosophers". I took a fair portion of this water and kept it. Then I went back with the old man to the Rosicrucian hermits. They showed me a small furnace, a pot, and a glass vessel which they used to cook the three salts, dissolve things in them, and reconstitute them to make excellent medicines. The old man spoke to me as follows: "Now you have seen all of Nature, all in all. Now you have proved that God has given to every single body a masculine seed to preserve it and also a feminine seed. The feminine seed takes its origin, preservation, and nourishment from this plenteous fountain. You have seen and understood how all things return to their first matter --- feminine seed to its spring, and masculine seed to another body --- and are then led back after their separation. And you have seen how they come together again in constant circulation. Behold, you now have a true compendium for investigating nature. You now have a laboratory in which all threes things can operate by themselves, and in which animal, vegetable, and metal objects can dissolve. Now you can work in such a way that you will please Nature and win honor for yourself".

Finis.

A Word from the Publisher

Thank you for purchasing this small work from The R.A.M.S. Library of Alchemy. During his lifetime, Hans Nintzel was dedicated to the identification, acquisition, study, retyping and, when necessary, translation of what he considered to be the most important known works on Alchemy. Hans was assisted by his sparse network of fellow Alchemists, all members of the Restorers of Alchemical Manuscripts Society (R.A.M.S.). I was an active member of R.A.M.S.

My goal is to publish all of the works originally made available through R.A.M.S. as photocopies. To facilitate this, I have chosen to have the books professionally printed. I also have a few titles that I intend to add to the original R.A.M.S. Library, selected by strict criteria established by Hans.

If you have a work on Alchemy that you believe should be a part of the R.A.M.S. Library, please contact me through R.A.M.S. Publishing Company.

Philip N. Wheeler